Army of the Sky

Studies in Modern European History

Frank J. Coppa
General Editor

Vol. 68

PETER LANG
New York • Washington, D.C./Baltimore • Bern
Frankfurt • Berlin • Brussels • Vienna • Oxford

Gregory Vitarbo

Army of the Sky

Russian Military Aviation before the Great War, 1904–1914

PETER LANG
New York • Washington, D.C./Baltimore • Bern
Frankfurt • Berlin • Brussels • Vienna • Oxford

Library of Congress Cataloging-in-Publication Data

Vitarbo, Gregory.
Army of the sky: Russian military aviation before
the Great War, 1904–1914 / Gregory Vitarbo.
pages;. cm. — (Studies in modern European history; vol. 68)
Includes bibliographical references and index.
1. Russia. Armiia—Aviation—History—20th century. 2. Aeronautics,
Military—Russia—History—20th century. 3. Air forces—Russia—History—
20th century. 4. Russia—History—1904–1914. I. Title.
II. Series: Studies in modern European history; v. 68.
UG635.R8V58 358.400947—dc23 2011043815
ISBN 978-1-4331-1490-8 (hardcover)
ISBN 978-1-4539-0266-0 (e-book)
ISSN 0893-6897

Bibliographic information published by **Die Deutsche Nationalbibliothek**.
Die Deutsche Nationalbibliothek lists this publication in the "Deutsche
Nationalbibliografie"; detailed bibliographic data is available
on the Internet at http://dnb.d-nb.de/.

29 Broadway, 18th floor, New York, NY 10006
www.peterlang.com

To my family, for all of their love and support,

And to Jill, a dear friend always

| Contents

| Acknowledgments

It would be impossible to name and properly thank all of the many people and institutions that helped me in ways large and small during the evolution of this project from hazy idea to wordy dissertation to finished book. One must eat in order to write, so credit is perhaps due first to my various sources of material support over the years. At the University of Michigan the CIC and Rackham Merit Fellowships underwrote the bulk of my time there, while teaching assistantships through the Department of History and the Center for Russian and East European Studies offered additional support as well as valuable teaching experience. The Mayer Travelling Fellowship provided critical funding for a year of archival research in Moscow. Survival would not have been possible without assistance from my incredibly generous parents as well. Visiting assistant professorships at Franklin & Marshall College and the University of Nevada-Las Vegas allowed me the luxury of continuing to teach and write. At Meredith a course release and later semester sabbatical granted by the Professional Development Committee gave me invaluable time and freedom to continue reflecting upon and revising the manuscript as I went about my usual day job of teaching and advising. My department and dean, Garry Walton, provided additional support for preparation of the manuscript.

Along the way I have met many incredible and supportive people, some of whom may not remember or even be aware of just how helpful they were, and thus might even be surprised about their mention here. The members of my dissertation committee—Julia Adams, John Shy, Ronald Suny, and Jane Burbank—were all incredibly generous in their patience, advice, assistance, and intellectual engagement. Special thanks must go to the chair of my committee and academic adviser, William G. Rosenberg. Throughout my years of study with him, he provided a model of intellectual achievement, professional integrity, and personal empathy that I could only hope to match. His influence upon my growth as a scholar, teacher, and person has been immeasureable. Yet dissertations are not completed without the support of equally miserable fellow-travellers, so I also wish to recognize my graduate school comrades, among them Paul Werth, Mary Cavender, David Althoen, Katherine Pence, and Anna Kuxhausen, who helped me during and beyond my years at Michigan. At research stints at the Hoover Institution, Bakhmeteff Archive of Columbia University, and New York Public Library the staffs were invariably helpful and accommodating. During my stay in Russia the staffs of the Russian State Military History Archive [RGVIA] and the Department of Military Literature of the Russian State Library were also unfailingly professional, gracious, and helpful. Irina Karapetiants and her staff at the Russian State University of the Humanities provided necessary assistance in navigating bureaucratic life in post-soviet Russia, while Sofia Ivanovna looked after me and my fellow dormitory ex-patriates with motherly concern.

Once home, Von Hadesty was generous enough to share draft copies of his work with me, while Mark von Hagen was kind enough to offer advice on a summer afternoon. During my sojourns in Lancaster and Las Vegas the faculty at each institution made me feel truly welcome and valued; Maria Mitchell, Joseph Fry, and Collin Loader stand out in this regard. At various conferences, meetings, and informal discussions, numerous colleagues, among them Dominick Lieven, Melissa Stockdale, Christine Ruane, Josh Sanborn, Eric Lohr, Peter Holquist, David Schimmelpeninck, John Steinberg, Bruce Menning, and Scott Palmer offered constructive commentary and advice that greatly improved the manuscript. Fellow faculty members in the Department of History and Political Science at Meredith College—Michael Novak, Daniel Fountain, William Price, Clyde Frazier, Hilary Smith, Jeffrey Martinson, Barbara True-Weber, Carolyn Happer—have been incredibly helpful and supportive in so many ways over the years; I have been truly lucky to be surrounded by colleagues who are committed teachers, gifted scholars, and caring, generous people. All of them provided invaluable insights and suggestions. Many thanks also to other colleagues, including Betty Webb for being a wonderful mentor, Diana Davis and Betsy Stewart for enabling daily life in our department to continue, and Robin Baneth for providing timely and invaluable help in the final stages of preparing the manuscript. My editors at Peter Lang, Caitlin Lavelle and Jackie Pavlovic, were patient, supportive, and enormously helpful throughout this process.

As noted, I am sure that I have forgotten to mention others who have contributed to this book in various ways, and for that I sincerely apologize. Moreover, despite all of this invaluable advice and assistance, any remaining failings of the book remain my fault alone.

Grateful acknowledgment is made to Brill Publishers for permission to reprint portions of the following material:

Gregory Vitarbo, "Military Aviation, National Identity, and the Imperatives of Modernity in Late Imperial Russia." In *The Military and Society in Russia, 1450-1917*, edited by Marshall Poe and Eric Lohr, 273-92. Boston: Brill Publishers, 2002.

Introduction

In the last decade of Imperial Russia the challenge of "modernization" took on new urgency. Modernization included such processes as the growth of mass representative politics and complex bureaucratic structures, the expansion of the scope and prerogatives of the state, technological advancement, and increasing socio-economic differentiation.The tangible manifestations of such processes—rapid industrialization, the growth of professional middle and industrial working classes, the spread of nationalism, the beginnings of mass political mobilization, cultural ferment—characterized a period of rapid, destabilizing change. Russian thinkers, politicians, and officials devised and modified conceptual and analytical models through which to interpret these processes. In turn, these models helped direct the evolution of such processes by contributing to the framing of governmental and public policy. Moreover, in addition to their own social and cultural heritage, Russians could seize upon the prior historical experience of the self-consciously "modern" Western Europe nations as an archetype through which to perceive, explain, and pass judgment upon the changes affecting Russia.

The dynamics of modernization took on a unique cast within the Russian Imperial army through the period from 1905–1914. As both the primary instrument and prop of tsarist power, the army occupied a crucial place, structurally and ideologically, within the edifice of the regime. By virtue of the demands placed upon it, the scope of the army's interaction with society was arguably broader than that of any other state institution. From conscription to repression to local administration, some form of military experience was shared, and hence in part shaped, by nearly every tsarist subject. The Imperial officer corps itself cultivated long-standing traditions of its prominent role and status, the ideal of state service which underpinned this status, and the deeper link between tsar and nobility symbolized by such service. Yet despite its apparently deep-seated affiliations to the tenets of the old order, the officer corps was unavoidably caught up in the contemporary political, social, and cultural flux that was challenging those very tenets.

Further, any study of the military in the last years of the Empire must contend with the inescapable shadow of World War I, the Russian Revolution, and the civil war. The upheaval of 1914–1921 has inevitably exercised a powerful influence upon memoirists and historians alike. This influence is frequently manifested in a "trial and judgment" model of analysis that holds the Russian army to an unforgiving paradigm of modernization and finds it wanting. The leitmotif of such work is the adjudged failure of the army to master objective challenges of technological advancement and military reform and the subsequent price of such failure—defeat in the Great War and the demise of the Russian Empire. This analytical orientation is to an extent understandable. The Russian officer corps in the early twentieth century was not engaged in merely philosophical debates regarding Russia's pride, place of importance, and cultural superiority vis-à-vis the West. Instead, it was in the midst of an increasingly feverish arms race and worriedly anticipating the outbreak of a major conflict, where defeat might mean more than mere humiliation or an anguished re-examination of "Russia."

Yet the impulse to assign blame frequently leads to a failure to appreciate sufficiently the context, parameters, and set of choices in which contemporaries operated. Military modernization was necessarily a complex and multi-faceted process. At one level it involved efforts to acquire new technology and utilize foreign doctrines and institutional models. The requisite policies and practices were more or less consciously emulated as "modern" or "Western", or at least allowing Russia to compete with the West. These efforts, however, must be placed within the broader context of domestic blueprints of modernization largely external to the armyand its control, but which nevertheless impinged upon its mission. Such projects, themselves envisaged as "modern" by their sponsors, included not only state-sponsored industrialization, but also such efforts as building a more coherent administrative structure for the Imperial polity. In turn, as perceptive Russian statesmen and thinkers had long ago learned, both artifacts of technology and programs of modernization imported from abroad were themselves underpinned by exogenous cultural values and historical experiences.

Further, if modernization was a reciprocal and dynamic process of negotiation, the terms of this exchange were mediated fundamentally by the culture of the tsarist officer community. By culture I mean that body of values, assumptions, and practices that structured how officers perceived, defined, and articulated their role, function, and identity. This set of values derived from their collective experience served as the starting point for any interpretation and appraisal of programs of modernization. Thus, one must pay particular attention to the unique features of this military community. The officer corps was an organization with specific functions and internally generated conventions and norms, but institutionally sensitive as well to international standards of military performance. At the same time, it was a part of the tsarist state apparatus whose relations with other parts of this apparatus, from the high sphere of the central ministries to provincial police officials, were varied, complex, and often contentious. Lastly, it was a community conditioned by broader links to prevailing social structures and cultural patterns in the tsarist empire.

This work specifically addresses the development of military aviation in the period 1905–1914 as a case study through which to explore the dynamic relationship between technology, the imperatives of modernization, and the culture of the Russian Imperial officer corps. The airplane was emblematic of the dilemma of modernization, for both army and state, in the last years of the Russian empire. It at once presented both a serious challenge and a tantalizing opportunity; it sharply exposed the limitations of Russia's economic, technological, and infrastructural development while simultaneously offering a means to rapidly overcome them; it provided a means to demonstrate and assert Russia's achievements, pride, and place while also giving rise to fears of the penalties of backwardness with the stakes of modernization now risen, literally, dramatically higher. The airplane thus offered a potent symbol around which definitions and visions of what modernization should mean for Russia could be contested. The advent of the airplane also called into being the need for a requisite cadre of trained personnel: the flyers and mechanics, the priests and acolytes, who would serve this new idol of technology. This task involved the creation and assimilation within the army of an almost entirely new profession of arms and a unique society of officers. In this way the airplane was the harbinger not just of a new age of technology, but of the birth of a culture of aviation within the Russian army, one that would reach its zenith of expression under the Soviet regime.

Given this symbolic power of the airplane, traditional indictments of the Russian military community take on heightened form in regard to aviation. They entail not only judgments of military performance in the air, but a broader critique of the inability of the tsarist army and state to confront the military, economic, and political challenges besetting the Empire. In contrast, the successes of aviation amongst various European powers of the period, particularly during World War I, offer a vivid comparative standard. This critical stance has again been exaggerated by the aftermath of the Russian Revolution. Instead of the broader, evolutionary continuity with which to judge the development of aviation in the United States and Western Europe, there is the seemingly stark contrast between the lame, feeble efforts of tsarist aviation and the large devotion of priorities, resources, and propaganda that surrounded aviation under the Soviet regime. In this view, the material backwardness and cultural failings of the tsarist military were graphically exposed, even before the debacle in war, by its response to the airplane. Soviet literature is united in painting a picture in which the tsarist officials and senior army officers failed to adequately adapt to the airplane and cultivate its "necessary" development. Even works which praise the efforts of early Russian fliers, inventors, and aviation enterprises stress the difficulties they labored under.[1] The best example of such work is the exhaustive study by P. Duz', *Istoriia vozdukhoplavaniia i aviatsii v Rossii* (*History of Aeronautics and Aviation in Russia*).[2] The composite picture, which remains largely the same in post-Soviet accounts, thus presents a sharp dichotomy. In the front lines of Russia's encounter with the airplane was a budding, vibrant culture of aviation professionalism, uniting a small group of "progressive" army officers with civilian inventors and enthusiasts. However, their efforts were ultimately stifled by the short-sightedness and inertia which generally prevailed both at the levels of command and the line, within the officer corps as a whole and even the aviation service itself. In turn, the general economic, industrial, and infrastructural backwardness fostered by the tsarist political and social order presented fundamental obstacles beyond the control of the officers of the air services. Hence, a visionary aviation culture was unable to flower fully until after the October Revolution and the Bolshevik commitment to modernity, symbolized by the regime's idolization of the airplane.

While somewhat more sympathetic, the small amount of Western work on the topic also treats Imperial military aviation largely as a preparatory footnote to the impressive aviation tradition of the Soviet armed forces. Such works generally focus upon the large-scale features of economic backwardness—a small industrial base, poor infrastructure, financial constraints—which hindered the development of Imperial military aviation and especially a domestic aviation industry.[3] These works also make mention of the energetic and fruitful activity of a handful of committed and innovative enthusiasts. Yet, like Soviet scholars, they depict such visionaries as largely alone and generally unheeded, like straws against the wind, amongst the general milieu of backwardness and obscurantism that characterized the late Imperial army.

However, in his history of Soviet aviation Robert Kilmarx does emphasize the debt owed by the Red Air Force to its tsarist predecessor—a debt that the Soviets were only too happy to minimize—in terms of inherited experience, techniques, methods, and theoretical knowledge.[4] Recent work on Russian Imperial aviation likewise presents a more nuanced picture. Scott W. Palmer's exploration of the prominent place and role of aeronautics and aviation in Russian popular culture throughout the tsarist and Soviet periods stands out.[5] He evocatively demonstrates that a tradition of aviation possessed long-cultivated and deep-seated roots in the Russian national memory. In a related piece he argues that in the years before World War I the airplane acted as a powerful symbol for the Russian public of the ability to overcome Russia's chronic "cultural stagnation and historical backwardness." Aviation thus offered Russians an example of strength and "a means of redefining their national identity," an identity pointedly contrasted to the obsolescent political and social order promoted by the Tsarist regime.[6] Likewise, Von Hardesty also chronicles the numerous achievements of Russian military and civilian aviators, scientists, inventors, and aeroclub enthusiasts in the years before and during World War I. He points to the renowned aircraft designer Igor Sikorsky, for example, as a glowing symbol of a Russian talents and initiative in this sphere.[7] Noting that it was long official Soviet policy to downplay the accomplishments of Imperial aviation, Hardesty instead argues that "one is struck with the continuity between the tsarist and communist periods" in terms of perceived challenges and responses to aviation, claiming that both periods evinced a common style "which could be described as distinctively 'Russian.'"[8] He thus stresses the numerous similarities—in philosophies, methods, and cadres—shared by Imperial and Soviet aviation policies.

While offering invaluable insights regarding the general development of Imperial aviation, however, neither Hardesty nor Palmer focus in depth upon strictly military aviation, which dwarfed civil aviation in this period. In turn, Palmer's analysis draws heavily upon the traditional model of a moribund autocracy pitted against an increasingly assertive public over the imperatives of modernization and the future course of the nation. Existing scholarship on Russian Imperial aviation thus offers both strident indictments of the army's failure to master the challenges of modernization, as well as indications that such judgments deserve a reappraisal.

This study provides such a reappraisal through a comprehensive and in-depth examination of the tsarist aviation service before the Great War. Limiting this study to the years before the outbreak of conflict has several advantages in terms of analytical and comparative context. It helps avoid the dangers of the aforementioned posture of judgment and the resulting tendency to view Russian defeat both in the air and on the ground as somehow preordained. It further allows a more fruitful application of an appropriate comparative framework, one that distinguishes between the pre-war period and the experience of the war itself, when the stalemate of the trenches dramatically increased the attention and resources devoted to aviation and the resulting pace of its development. Before the exploits of the Great War's aces crystallized the popular image of early aerial combat, aviation was a novel and dodgy business. All of the European militaries grappled with the challenge of the airplane with varying levels of commitment and success; it was a highly competitive process of trial and often deadly error, where relevant knowledge and useful experience were scarce, standards and technologies evolved rapidly, and past accomplishments were no guarantee of future success. France was widely recognized as the leading military aviation power before 1914, yet it was the Germans who arguably did the most to revolutionize both aircraft technology and tactics under the pressure of war. In 1914 the Russian air force could boast second place among European armies in terms of number of aircraft, but the demands of war quickly exposed such claims as hollow.

Related, within this comparative context, taking August 1914 as an endpoint allows analysis of the personnel and culture of the pre-war tsarist air force on its own terms, before the demands of the conflict fundamentally transformed both its size and character. Although it would grow rapidly, the aviation service was quite small before 1914. Due to the resulting paucity of positions of command, it was primarily populated by exactly those junior officers—lieutenants, captains, staff-captains—that often prove so elusive in studies of the officer corps. Yet while the aviation service attracted officer volunteers from all the various branches of the army, it was not considered an elite division. Sociologically and culturally, it was therefore comprised of a broad, uniquely representative cross section of tsarist officers. As a result, the community of military aviators would draw upon established models and conceptions of duty, service, and identity while simultaneously elaborating new values, roles, and paradigms of service distinct to aviation.

Thus, if from the vantage point of 1917 the Russian Imperial air force had itself been significantly transformed, and the army and the regime it served overwhelmed, by the demands of the Great War, it was in the pre-war years that the foundation of this service was established. This involved organizational infrastructure, training and staffing policies, a core cadre of experienced aviators and command personnel, and, equally importantly, a nascent institutional memory and culture. It was this foundation, limited though it was, that would support the dramatic expansion of the wartime air service and also serve as a bridge to the Bolshevik era. While numerous officers joined the emigration, Imperial aviation veterans, soldiers and officers alike, alongside various pre-war inventors, technicians, and scientists, played an important part in creating and nurturing the Red Air Force.

A set of questions therefore remains. Were the demands of the airplane and its aviators indeed irreconcilable with a prevailing culture of backwardness, both within the army and the regime at large? Were there elements of tsarist military culture that served to promote aviation, as testified to by the existence of officer aviation enthusiasts themselves? Was aviation in some ways destabilizing, exacerbating cleavages and fractures within the officer community, or was it perhaps unifying and integrative? To what extent did this culture of aviation service draw upon broader attitudes and influences, both from Imperial society and culture and from the international experience of aviation? Lastly, did the perceptions of modernization and alleged dissidence of tsarist aviators match the Bolshevik vision of modernity later ascribed to their efforts, both by the Soviet regime and by those former tsarist officers who served in the Red air forces?

Fortunately, there is a large and diverse literature to draw upon in addressing such questions. In addition to the specific studies of the tsarist air forces cited above, there is also the rich body of work on the air forces of Europe, in particular the English, French, and German. Further, an increasing amount of work has been devoted to exploring tsarist officer culture and modernization in the Imperial army.[9] An admirable model in this regard is provided by John Keep's excellent social history of tsarist soldiers and officers of the pre-1874 army.[10] In a broader comparative frame, there are the socio-cultural histories of various European officer corps, among them excellent works by Demeter, Deak, and others.[11] Along with Keep's work, they offer a model for considering the sociological, cultural, and political experience of officers in sum. Lastly, the use of overtly sociological and cultural approaches to studying issues of warfare has become increasingly popular in past years; John Shy notes that the value of such approaches lies in their broad, inclusive, inter-disciplinary potential.[12]

In the realm of theory, the multi-faceted experiences of the officers of the Imperial aviation service can be examined, for example, through the prism of militarism. Although he is most well-known for articulating an ideal model of civil-military relations through his archetype of a "professional" soldier, Alfred Vagts also discussed another variant of militarism.[13] This entails attitudes and frames of thought that interfere not with politics and civil structures, but that place institutional values, interests, or inertia before what he defines as the chief task of any army: scientifically analyzing and preparing for the next war. In his words, militarism "presents a vast array of customs, interests, prestige, actions and thought associated with armies and wars and yet transcending true military purposes... militarism displays the qualities of caste and cult, authority and belief."[14] While at first glance this definition may seem to draw upon the value judgments of the historiography cited above, it can be used to explore in greater detail how exactly Russian officers conceptualized issues of duty, service, and competence.

William Fuller's treatment of military professionalism in the late Imperial army provides a complementary analytical frame. Instead of empirical results, Fuller emphasizes the creation and maintenance of a self-conscious professional identity, cultivated through rigorous selection, training, and standards of performance and behavior. Within this context he stresses five criteria, again to be measured less by "success" than by commitment to these ideals: a specialized set of knowledge and skills; the insistence on the upgrading of standards of performance; a strong sense of group identity; a recognition of the special interests of the military; and autonomy over criteria of admission and promotion.[15] While employing these criteria, Fuller avoids the extremes of a rigid and reductionist application of them to the understanding of professionalism on a daily level. Like modernization itself, professionalism was more a matter of process, a state of mind, than some objective and static result.

Lastly, the words and actions of the aviation officers themselves—as represented in official reports and correspondence, orders and regulations, contemporary interviews and publications, press accounts and memoirs—provide the most compelling sources for illustrating their attitudes, values, aspirations, and self-perceptions. They allow one to examine the linkages of culture and modernization within the tsarist aviation community through various lenses: broader valuations of technology; particular hopes and fears inspired by the airplane; specific projects drafted to develop military aviation; institutional adaptations made in response to the airplane; relationships and conflicts amongst senior officers of the aviation institutions; policies of recruitment and training for the cadre of military pilots; practices to cultivate a requisite service ethos for such officers; ceremonies of both celebration and mourning. Through these projects, debates, practices, and rituals, such officers addressed not only the priorities and potential of military aviation, but such issues as the proper role of the army, its ideal nature and composition, its relationship to the society and culture around it, and conceptions of empire, state, and nation as they related to notions of service and duty.

By exploring the collective experience of this community of officer aviators, a community at the forefront of the tsarist army's ongoing confrontation with the imperatives of military modernization, this study will thus provide a fuller, richer picture of the Russian air force in its formative years before the Great War. Examining this experience within the milieu of the tsarist officer community as a whole, as well as within the broader comparative context of the pan-European development of military aviation, demonstrates how this community articulated a uniquely Russian experience of military modernization. Lastly, by illustrating the connections between this community and prevailing social, economic, political, and cultural currents and dynamics outside the realm of the army itself, this study can provide a window into the broader meanings, possibilities, and limits of modernization and modernity in the last decade of the Russian Empire.

1| Technology, Community, and Reform in the Russian Imperial Army

On December 17, 1903, two enterprising American brothers named Orville and Wilbur Wright made their first powered flight in an aircraft of their own invention in Kitty Hawk, North Carolina.[1] The flight was quite short, the brothers little known, and only in time would this date be universally recognized as a true turning point in the history of humanity's march of progress. Almost two months later, on February 8, 1904, halfway around the world, Japanese naval forces launched a surprise attack on the Russian Far East fleet at its base at Port Arthur. The attack was the culmination of over a decade of tension between the two empires over Russian expansion into Manchuria and Korea.[2] The subsequent war seemed to presage a far more dramatic advance in the exercise of military power and the lethal application of technology than the brief flight of two obscure Americans.

With battlefield defeat compounded by social rebellion, political turmoil, and military repression, the years 1904–06 were traumatic ones for the tsarist regime and its armed forces. Only with the ebbing of unrest within the army after July of 1906, followed by the gradual pacification of the Empire at large, would some sense of calm return to the army. Peacetime routine provided only the semblance of normalcy, however. Far more than a strictly military defeat, the experience of the war profoundly struck at the officer community's sense of self. Looking inward, the army itself had been rent by mutinies amongst soldiers and, perhaps more disturbingly, disloyalty among comrades in arms, as numerous reserve and even some active junior officers participated in revolutionary activity.[3] Looking outward, many officers saw a regime which had seemingly abandoned them on the field, while the public and nation as a whole had questioned and then violently repudiated their service and sacrifice. The resulting dismay of such officers was heightened by a palpable and lingering sense of humiliation, replete with a racist undercurrent of cultured Europeans being bested by barbaric orientals.

As the army and its achievements were so intimately and deeply identified with Imperial greatness and Russia's place in Europe, the triptych of war/revolution/repression thus had a profound, yet disparate, influence upon the officer corps' sense of its role and place within the Russian Empire. As was the case following the Crimean War, a widespread reaction among the

Empire's military elites was an anxious analysis of Russia's military "backwardness" and recognition of the need to reform and modernize the armed forces.[4] However, "reform" was inherently problematic, with no "objective" understanding of ills or remedies;[5] the result was not consensus but a plethora of diagnoses, proposals, and cliques.[6] Some deficiencies were relatively easy to recognize and address, while more far-reaching proposals generated enormous controversy. Various scholars have shed light upon the resulting extraordinary intellectual ferment within the tsarist officer community regarding issues of strategy, doctrine, war planning, and tactics.[7] Further, the larger context of such debates was inevitably informed by the dynamic social, political, and cultural milieu of late Imperial Russia as a whole. A host of ostensibly non-military issues would ultimately affect the army's efforts to reform and modernize almost as much, if not so overtly, as direct challenges like financial constraints and investment priorities. Lastly, the Russian officer corps was by no means an independent agent in determining the fate of the army. Instead, it necessarily operated within the larger realm of the relentless pace of military innovation and competition amongst the armed forces of the European powers, all of them to an extent hostage to the changing nature of warfare itself.

The question of how best to respond to transformations global in scope and yet with such profound and immediate implications for the Russian army thus provoked replies that addressed fundamental and enduring themes. These themes involved the familiar juxtapositions of tradition and imitation, of backwardness and progress, of Russian uniqueness and international pressures, of Russia and the West. In retrospect, however, all prescriptions for reform within the tsarist officer community, from the comprehensive to the cosmetic, revolved around two fundamental issues: how the Imperial army would wage war in the future, and the nature of the army that would wage it.

Waging War: Technology and Morale

Technology had long occupied an ambivalent place in Russian military thought. In the context of a relatively underdeveloped industrial base, financial constraints, a large pool of rural, poorly educated peasants, and a proudly cherished historical legacy of military success, masses of soldiers seemed to Russia's military elites to be her best and most plentiful resource.[8] In the realm of strategy, virtues were made of this necessity, hence the army's traditional emphasis upon geographic space, manpower, and casualties to surrender as compensation for perceived technological superiority amongst Russia's enemies. Backwardness in this sense could be seen as a military advantage. N. P. Mikhnevich, Commandant of the General Staff Academy from 1904–07, argued in his writings that Russia's agrarian nature and comparative backwardness would spare it from the wartime upheavals which would disrupt more complex Western economies and societies; as Menning notes, this mirrored the nineteenth century Slavophile belief that Russia's backwardness was actually virtue.[9] In like fashion, tsarist officers could point proudly to the unique virtues enshrined in what John Bushnell calls "the cult of the soldier", which praised the incredible stolidity, endurance, and courage of the Russian peasant soldier, envied by nearly every army in the world.[10] As such, in Russian military thought, the decisive role in battle had traditionally been granted to the infantry, with primary emphasis placed upon the power of morale.[11] The most influential Russian tactical theorist in the years before and after the Russo-Japanese War, General M. I. Dragomirov, stressed above all the decisive role of the will and courage of the individual soldier; generally resistant to technological change, Dragomirov insisted that in battle weaponry was subordinate to morale, firepower secondary to cold steel.[12]

The experience of the Russo-Japanese War had an uneven impact upon such attitudes. Few officers who had served in Manchuria could ignore the increasing scale, scope, complexity, and technological orientation of modern warfare, the latter manifested in such weapons as quick-firing artillery and the machine gun and related technologies such as railroad, telegraph, and automobile.[13] Such developments wrecked havoc upon traditional military conceptions of time and space, while not coincidentally rendering modern combat particularly gruesome and lethal.[14] Yet most Russian officers, as well as their European counterparts, a number of whom had witnessed the war first-hand as military attaches and observers, failed to grasp the full scope and implications of this fundamental revolution in warfare. While there were indeed tsarist officers who advocated a fundamental reassessment of the role of technology and the importance of defense in Russian military practice, other officers concluded that these same factors made offensive operations and high morale even *more* vital to achieving victory. Convincing evidence was provided by the Japanese themselves, who by consistently taking the initiative achieved victory against difficult odds and despite severe casualties.

When compared to the other armies of Europe, the enduring emphasis upon the role of the infantry and the power of morale was not in itself unique. However, the uniqueness of the Russian context could push European ideas to extremes; as Peter Kenez argues, "in Western Europe belief in the importance of the will did not hinder the development of modern technology, but for Russia, which had limited technological capabilities, the same thinking became a justification of the status quo and a hindrance to understanding the real causes of the weakness of the Russian army."[15] This point is somewhat overdrawn. Walter Pintner argues that up until the late nineteenth century the Russian army was basically successful in "modernizing," if the issue is defined as acquiring and deploying the newest military technology.[16] While material shortcomings persisted after 1905, the army in the immediate years before 1914 was taking aggressive measures to improve and update its weaponry and equipment;[17] financial constraints and government penury presented far more difficulties in this regard than any systematic aversion to technological modernization.[18] Even so, the Small Program of 1910 and the Large Program of 1914 were to provide massive expenditures for the procurement of weapons and related technologies. There were in addition a spate of administrative and organizational reforms in the years following the war, and field regulations and tactical directives were consistently re-written to take into account new weapons and their tactical applications.[19]

However, Kenez is correct in suggesting that a marked institutional orientation toward human and moral resources formed the parameters of tsarist officers' appraisals of technology. There was a prevailing realization that the Empire could not compete militarily on the basis of technology alone. Nor was trying to do so, some believed, necessarily desirable, in light of the perceived price paid by European armies for the industrialization that fed their arsenals: labor unrest, malnourished and politically unreliable recruits, and societies obsessed with crass materialism instead of noble martial values. As Pintner notes, "perhaps only in Russia, because of the sensitive nature of the broader issues of Russia and Europe," could emphasis upon either modern technology or the power of morale be the basis of such deep divisions within the Tsarist officer corps.[20] Such ambivalence coalesced in the debate between the "Nationalists" and the "Westernizers." As Menning phrases it, "one side [the Westernizers] preached the merits of military modernization, whatever the inspiration; the other [the Nationalists] trumpeted the necessity to search the immediate and more remote past to retain harmony with Russia's true national lines of military development."[21] While the respective positions were not mutually exclusive, they represented two distinct paths of reconciling the army's history with the challenges of the future.

Not surprisingly, technology was one of the flashpoints of this debate, as it was perhaps the most tangible manifestation of the pressures of modernization, investing at once issues of tactics, strategy, and philosophy. Mirroring the contours of the larger debate between the two camps, the dialogue was somewhat muddled, but there were still discernible differences, and they were inevitably framed around the traditional theme of the power of morale. For example, discussions about the effects of modern artillery invoked such disparate reference points as Imperial military traditions and glories and the modern social science of psychology. Numerous commentators pointed to deeper sources of morale that helped soldiers endure the hell of modern artillery fire: an awareness and appreciation of larger military goals and national greatness, a supportive public, a unified nation. A regular theme in the tsarist military press, and frequently cited as perhaps the key element in Japan's triumph in the recent war, was the belief that patriotic unity had given the Japanese soldier intrepid determination and an unshakeable belief in victory. Such recognition in itself was quite flexible. Traditionalists as well as reformers could assert that the problem lay not with the Russian soldiery; it was instead society and its anti-military, unpatriotic attitudes which needed to be thoroughly reformed. A contributor to *Voennyi sbornik* in 1911 complained that in Japan, in contrast to the Empire, "the government and people worked consciously and in complete agreement on preparation for a national war"; as was custom, he explicitly blamed the Russian intelligentsia.[22] Nevertheless, many officers recognized that a fundamental hindrance to any reform, technological or otherwise, was the forcibly conscripted peasant-soldier himself, a problem that attracted widespread attention in the official military press of the period and subsequent memoirs.[23]

As such, many believed that the key to modernizing the army was the creation of a motivated, patriotic, civically conscious soldier in service to a united, equally patriotic nation. This ideology of a united army and society, "the nation in arms", was believed to be possessed by advanced powers such as Germany, France, and Japan. Yet it was a quite radical prescription for Russia itself, calling for the Imperial army to construct this new citizen-soldier out of what was perceived as a mass of illiterate, fatalistic peasants in a society that was largely apathetic, and often hostile, to the military and its values.[24] Thus, the relationship of technology, morale, and the Russian soldier involved far more than purely "military" concerns. As in other European nations, the Russian variant of the "nation in arms" was theoretically founded upon the broader adherence to concepts of nationalism and patriotism within society as a whole. Its proponents recognized that the challenge of constructing the new citizen-soldier was not the army's alone, but would necessarily involve the cooperation of the government and populace as well. Given the perception of continuing hostility to both the military and the autocracy among the educated classes, cultivating this support would be a daunting task.

Culture and Community

As the technology/morale debate makes clear, one of the by-products of the Russo-Japanese War and the aftermath of revolution and repression was to spread appreciation of larger currents of change throughout the officer community as a whole. Even the most committed obscurantist could not help but acknowledge the momentous political, social, and cultural transformations at work in the Empire. Less committed obscurantists went further and recognized that such changes intimately affected the army and effectively shattered its self-cultivated image of an institution largely insulated from the outside world. Moreover, many committed to comprehensive reform necessarily looked to the officer corps itself as not just an agent but also an object of change. The unconquerable Russian peasant soldier formed only one half of "the union between officer and men" in the Imperial army. On the field of battle, it was individual genius, in lieu of any rigid doctrine or even common training, which conferred moral authority upon an officer and provided the key to victory through the seamless bond between him and his men. This emphasis upon inspired leadership certainly did not preclude tsarist officers, particularly those in the engineering and artillery services, from receiving first-rate educations and technical training and developing a professional ethos of service.[25] But like the emphasis upon morale for the soldiery, the stress upon individual genius and leadership constituted a broader cultural context which conditioned how tsarist officers conceived of their service and duty.

In one sense, this could have a positive effect, as these abstract ideals provided a sense of common purpose within an increasingly differentiated officer corps. Yet contemporaries and historians alike have blamed the failures of the late Imperial army and the ultimate roots of defeat in World War I upon the unique features of acculturation and socialization within the tsarist officer community. Both groups most often cite the striking lack of unity within this community itself. Officers were distinguished by "striking differences in outlook, social standing, educational attainments, and wealth."[26] These differences were exacerbated by cleavages based upon training, branch of arms, and geographic posting, such that "an officer's life, privileges, and conditions of service varied in strict dependence on the sort of unit in which he served."[27] Such conditions largely precluded any overarching sense of common identity within the officer community. Instead, numerous sub-cultures largely discrete and even somewhat hostile to each other were the axes of self-identification for most tsarist officers;[28] "constant rivalry, lack of understanding, and jealousy undoubtedly weakened the army."[29] Within this context, "it is understandable that officers possessed no sense of belonging to a common institution," and even the term "officer corps" was something of a misnomer, and it was rarely used by contemporaries.[30]

In turn, the few unifying cultural elements among Russian officers are generally identified as negative ones that actively inhibited military functionality and efficiency. Regarding the senior levels of the army, scholars emphasize institutionalized habits of inertia, the observance of strict seniority, personal enmity, a penchant for intrigue, the role of personal connections, and corruption as the culprits responsible for flawed policies, questionable priorities, and material deficiencies.[31] Worse, such vices allegedly filtered down the chain of command and permeated the officer community as a whole. At the level of the line, scholars often focus upon more exotic vices peculiar to the tsarist officer community: gambling, rampant alcoholism, womanizing, dilettantism. This is a well-established tradition that was born in the late Imperial period itself with Aleksandr Kuprin's biting satire *The Duel*, with its host of unsavory characters that populated a "typical" tsarist regiment,[32] and was carried on following the Revolution in the memoirs of Soviet as well as émigré officers like Gen. A. Denikin.[33] In John Bushnell's scathing indictment of the late Imperial officer community, he argues that such dubious social mores were reinforced by the pernicious effects of a regimental economy that socialized junior officers into a clerk's mentality, one marked by passivity, idleness, and indifference to strictly military duties.[34] Thus, "tsarist officers sustained patterns of behavior incompatible with their military mission": their behavior on duty was "perversely unmilitary," their off-duty behavior "anomic."[35] Likewise, the emphasis upon inspired leadership frequently manifested itself in a widespread indifference to or distaste for rigorous study, which contemporaries and historians have cited as one of the more pernicious traits of the tsarist officer community. Such attitudes were the product of a profound anti-intellectualism, which denigrated the value of a common military doctrine, the application of scientific principles to the art of war, or even basic knowledge of current advances in military practice.[36]

Moreover, scholars assert that such corporate patterns of behavior were shared throughout the officer community, transcending its numerous divisions and cleavages.[37] However, such behaviors did not impart any positive sense of a common identity; they provided tsarist officers only "a unity of which they themselves could scarcely conceive."[38] Perhaps the clearest expression of such unity was a purely negative and often violent caste mentality directed towards a public perceived as irredeemably hostile to the army and all that it stood for.[39] Both Bushnell and William Fuller emphasize this concept of "negative corporatism." While citing some differences with Bushnell's conclusions, Fuller argues that "insofar as corporate identity or consciousness existed for the entire Russian corps at all, it existed *only* in relation to perceived menaces from the distrusted outside world;" this bestowed no more than a "fictive unity," based almost exclusively upon shared fear.[40]

Worse, such pathologies allegedly persisted despite undeniable currents of change within the officer community. There is ample statistical evidence of the increasingly “middle class” composition of the officer corps in this period; growing numbers of men of diverse backgrounds pursued the military careers that were no longer attracting nobles. While a long-term process, it accelerated after 1905. In his systematic illustration of the connections between the social composition, career patterns, and service ethos of the Imperial officer corps, Peter Kenez terms this “an important social transformation” of the officer corps in the years before World War I.[41] The army was thus an important vehicle of social mobility: for example, by 1912 almost half of the infantry officers were descendants of serfs, while by 1914, only 40% of infantry officers were of noble origin.[42] Yet like Bushnell, Kenez implies that the few shared cultural values of the officer community—a fundamental conservatism, political naivete, anti-intellectual bias, and mistrust of civilians—were largely successful both in assimilating its changing social composition and in deflecting pressures for modernization and reform.[43] Thus, the purely negative unity of the Russian military community was proof against any great disruption or transformation in its values or attitudes.

At the same time, the increasingly diverse social composition of the officer corps nevertheless provoked ambivalence and anxiety among officers of noble extraction. Even if one accepts that increasingly heterogeneous social elements were assimilated into the army with little change in the ethos of the officer corps, the army was not blessed with this hindsight. Thus, the regime and army leadership responded with a “blatant class policy” which attempted to restrict social mobility in numerous ways, such as disproportionately funding noble military schools.[44] Such efforts failed. While the senior ranks continued to be represented overwhelmingly by nobles,[45] the profession of officer in general became a middle class occupation.[46] Yet this gulf only exacerbated concerns over disunity and dissonance within the army.

For decades after the war and revolution, the picture of the late Imperial officer corps thus remained both depressing and stubbornly persistent: a discouraging mix of apathy, incompetence, and poor morale, united only by the “fictive unity” of hostility towards those who did not wear the uniform, yet more significantly divided by profoundly disparate realities. On the one hand were a few highly educated officers dealing with grand designs of reform and modernization within the ivory tower of the General Staff in St. Petersburg, alternatively dabbling at court and making the rounds of high society. On the other hand was the almost feudal regimental reality in the provinces, with the stereotypical image of the career mid-ranking officer stuck in the mud of Poland or Ukraine, carefully tending to his seniority, searching for escape (often in liquid form) from the mind-numbing boredom and mountain of paperwork, largely untouched by and even more disinterested in the debates that would ultimately decide his career and fate.

Given this social and cultural context, various scholars conclude that any comprehensive programs of military reform and modernization in the pre-war period were difficult to realize. Wildman describes an officer cadre increasingly marked by "professionalization based on education," but notes that this process was not the same as promotion based on merit. Instead, "the long-standing weakness of giving first consideration to seniority and connections over talent and knowledge remained untouched by reform" even after the Russo-Japanese War, often bred apathy and atrophy among officers.[47] As such, an increasingly competent cadre of officers who were unable to penetrate the malaise imposed by the culture of army service itself. Kenez likewise cites the "peculiar Russian illness" whereby a good many officers served not in the field but in schools, ministries, and staffs, yet generally received faster promotion and greater privileges, all of which led to resentment among line officers.[48] Bushnell notes that young officers who exhibited initiative, zeal, or the desire to improve their skills were met with suspicion, scorn, or incomprehension from colleagues and commanders alike.[49] Fuller asserts that "it might be even plausibly argued that a military profession as profession in late Imperial Russia simply did not exist."[50]

While not denying that pockets of professionalism and progressive thinking existed within the officer community, such scholars inevitably add the caveat that these qualities were highly localized in small and highly unrepresentative groups such as military engineers and senior policy makers of the War Ministry. Scholarly work which focuses specifically upon these groups, such as Fuller's examination of military lawyers and Matitiahu Mayzel's and John Steinberg's studies of the General Staff, tend to do so in part because these groups appear more homogenous, more cohesive, and more progressive, thus using the faceless and amorphous mass of the army itself as a foil.[51] The General Staff in particular prided itself upon being a rigorous meritocracy and professional military elite, quite distinct from the lumpen proletariat of line officers. Thus, it is more difficult to link conclusions regarding such elites groups to the culture of the officer corps as a whole.

Lastly, these often grim depictions of the Russian officer community, both material and cultural, are magnified when judged against contemporaries elsewhere in Europe. In this perspective, unlike the armies of France, Germany, or Britain, the Imperial army failed to acquire and develop adequately new weapons and related technology, it failed to elaborate the strategy and tactics to harness this technology, and in general it failed to cultivate a culture of competence and professionalism among its officers. In short, the tsarist officer corps was profoundly unprepared for the challenge of modern war, a fact made explicit by the disasters of World War I.

However, while not without merit, critiques such as that provided by Bushnell, Kenez, and Wildman tend to be somewhat anecdotal, with individual experiences generalized into something of a canon for the entire officer corps. As Fuller points out, "we should be very leery of accepting impressionistic observations as definitive statements about the group behavior of 35,000 people," especially as some of the evidence is itself contradictory.[52] Moreover, while he himself employs a model of "negative corporatism," Fuller also correctly points out that such negative critiques tend to overemphasize Russian distinctness. Frequent suicide and red tape, for example, were features of the German and French armies, respectively.[53] Judging from memoirs, the Germans, those models of military efficiency, do not seem to have had an aversion to alcohol either. English officers likewise had a notable affinity for sports and games of chance, befitting the gentlemen that many of them were. The French officer corps was rent by political, religious, and class divisions laid bare by the ugliness of the Dreyfuss affair. The institutional response of the officer corps—a closing of the ranks and a palpable distrust and smoldering hostility towards the liberal public and its political leadership—nicely fits the description of the purely negative caste mentality of the Russian officer corps. Lastly, other features of the Russian military which draw particular criticism—the importance of social and class considerations in career advancement, the disproportionate influence of Guards corps, a frequently ambivalent prioritization of strictly military concerns—were shared by other armies, notably the English and the Germans, whom are judged to have managed the tasks of modernization effectively.

Most importantly, such critiques of tsarist military culture generally do not analyze or largely cannot adequately account for those currents of professional or progressive behavior amongst the mass of officers, nor any pressures for change these officers may have generated. Bushnell concedes that his analysis tends to describe a system "with a built-in tendency toward equilibrium," and thus cannot account well for undeniable historical evolution. He seeks to escape this by arguing that "these elements of modernity, however, were set in a premodern framework"; improvements were in the details, while the underlying system of non-military behavior persisted.[54] Fuller also observes that Bushnell's model cannot account very well for manifestations of military professionalism.[55]

Yet as noted above, Bushnell and others themselves concede the existence and activity of reformist and professional officers, even if their ambitions were ultimately unrealized. Wildman describes increasingly educated, professional, yet frustrated officers, Kenez notes how line officers complained of the unjustified perquisites of Staff and Guards officers, all note the throngs of line officers who yearly flocked to take the General Staff entrance examination and escape a fate of misery in the line. Kenez also cites the numerous notable World War I generals, Denikin among them, who rose from humble backgrounds to the heights of the Russian army leadership.[56] Likewise, although somewhat murky and difficult to quantify, a thorough reading of the contemporary literature and periodicals, as well as subsequent memoirs and biographies, confirms the existence of positive sources of socialization and reformist efforts throughout the various branches and strata of the military community. Yet these scholars must account for such alleged anomalies of professionalism largely by referring to individual ability, desire, and the role of chance; such factors allowed these officers to escape to the General Staff or some other choice position before the culture of line service inevitably turned them into the drunken, apathetic caricatures of Kuprin's fictional regiment. However, the argument that all failures and dilettantes were products of tsarist military culture as a whole, while professionals and progressives flourished only in spite of it, seems somewhat difficult to accept.

As such, recent works have offered, if not quite a rehabilitation, then a more nuanced and complex portrait of the Russian officer community and its multi-faceted attempts at self-examination and improvement.[57] In different ways, they explore the challenges, patterns, and dynamics of reform and modernization in a long historical arc, while situating them in a broad comparative context. Moreover, they approach and examine the unique institution that was the tsarist officer corps in the same way, while further emphasizing that this was a community whose values and attitudes were conditioned by broader links to prevailing political institutions, social structures, and cultural practices throughout the Empire. Summarizing the perspective of much recent work, Denis Showalter identifies three key points regarding the willingness and ability of the Russian army to adapt to the challenges of reform: the autonomy of Russia's military experience, the army's rational actor pattern of behavior (not any cultural "defects"), and lastly, and "most shockingly revisionist", the essential competency of the Russian army, in the eyes of both potential allies and enemies.[58]

As such, recent works emphasize that tsarist military culture was more than a collection of dubious mores and salacious conduct; it manifested itself in deep-seated habits, values, and paradigms of service, duty, and loyalty as well. I would emphasize such issues as attitudes towards education and training, esteemed as much for their assumed role in socializing as for imparting knowledge and skill; the way power and authority were structured and exercised, as manifested in issues of discipline, punishment, arbitrariness, and formalism; the role of personal honor, even more than rank, as the locus of identity, competence, and authority; the available institutional and social channels to express criticism; the maintenance of institutional memory and the dissemination of knowledge gained in different theaters of war and frontier; and the role of gender in shaping constructions of hierarchy and discipline. Such values and habits were the sources of vices as well as virtues, spanning the spectrum from Vagt's retrograde aspects of militarism to Fuller's attributes of professionalism, nor did they guarantee "correct" diagnoses and solutions to the challenges of reform and modernization. However, they must be understood and appreciated in their own right.

This is especially true regarding the issue of military professionalism. Again, it is against the litmus test of international military professionalism that the tsarist officer corps usually fares so poorly, despite the fact that the Empire is usually perceived as having been saturated with military men and military values. Yet Fuller himself emphasizes that professionalism does not necessarily entail a "progressive" outlook, correct policy decisions, or particular wisdom, while he asserts that "professionalism and negative corporatism could and did coexist in the minds of Russian officers."[59] Indeed, many reformist officers of this period who embraced what might be termed the professional impulse discovered that it was fraught with its own tensions, contradictions, and multiple possibilities. Regarding the relationship of officers and soldiers, they argued that elevating the morale and performance of the men under their command would require more than inspiring words, feats of courage, or strokes of genius on the part of officers. Moral authority could only come from better education, training, discipline, and commitment, which would grant soldiers faith in the competence of their leaders and thus an unshakeable belief in victory. For example, an officer who sought an ideal "higher than a medieval knight" simultaneously advocated a comprehensive, scientifically formulated education for officers which would develop the moral element in all its manifestations, including the "higher" feelings such as love of the fatherland.[60] Other vocal advocates of professionalism included the Zealots of Military Knowledge, the "Young Turk" faction at the General Staff and Academy, and the unofficial progressive military journals *Razvedchik* and *Voennyi golos*.[61] Yet it was difficult to spread a coherent gospel of professionalism in this fractured officer community, where many of the talented and ambitious were resigning their commissions due to difficult material conditions, low expectations of promotion, and poor morale.[62]

Moreover, advocates of professionalism realized it could not prosper in a vacuum. As the debates regarding morale and the soldiery made clear, military matters were inextricably intertwined with broader issues. To return to the issue of the nation in arms, foreign military thinkers realized that this doctrine could only be cultivated within the framework of a dedicated, professional army which enjoyed the support of a united public.[63] Tsarist officers by and large realized this as well, yet their task was a difficult one. Wildman argues that the paradigm of the "nation in arms" was in fact fundamentally incompatible with the backward, preindustrial, hierarchically stratified society of the Russian Empire.[64] Moreover, particularly rankling was the distaste and even contempt of the educated classes for the army, attitudes which never failed to provoke a critical or bitter comment in the military press. Such attitudes made the larger consensus of motivations and goals implied by the "nation in arms" even more elusive, while simultaneously making it difficult to attract the best and brightest to a career in the officer corps. Even given all of the vaunted power of the army, officers often felt alien in their own country.

Lastly, despite the efforts of the Duma and a few military professionals, both the tsar and the government bureaucracy, including much of the army leadership, were for the most part fundamentally reluctant to give the "nation" a voice in even the more trivial matters of national policy making. The participatory role, if only on an emotional level, that the "nation in arms" paradigm implied for the public as a whole was clearly beyond the pale. This suspicion applied to military servitors as well. As John Steinberg argues, even General Staff officers were ultimately unable to establish a true professional identity, and the autonomy that demanded, due to the prevailing political structure of a bureaucratic autocracy.[65] From another perspective, David Rich concludes that even successful cultivation of a professional impulse within the General Staff was thus inherently subversive of the traditional relationship of autocratic authority, state power, and the role of the army.[66]

The questions of how the Russian Imperial army would wage war in the future and the nature of the army, and in particular the officer corps, that would wage it were thus closely intertwined. Moreover, these questions required answers that in some way transcended the immediate sphere of the army to involve the public and society at large. It was perhaps this aspect of reform that proved so unsettling to many tsarist officers. It was increasingly difficult to couch even the seemingly objective and rigid demands of technological modernization narrowly; one could instead see a complex and almost infinite web of relationships between armament, soldiers, officers, public, nation, and regime. Acquiring and deploying new technology was if not easy than manageable. In contrast, effecting what some officers deemed as necessary transformations within the officer community itself, or in Imperial society and culture more broadly, was more elusive and problematic. And while Russian officers were acutely aware of the nature and stakes of these challenges given the Hobbesian world of military competition in the decade before World War I, even the most ardent advocates of reform and modernization emphasized the need to do so in a way that respected the history, traditions, and culture of the Russian Imperial army.

Aviation and the Officer Corps

In particular, the response of the Russian Imperial army to the advent of the airplane as a novel and ultimately revolutionary weapon provides unique insights into the dynamics of modernization and reform within the context of the tsarist officer community. As noted in the introduction, this response has been judged harshly by historians. Regarding the dual themes discussed above, Soviet scholars such as Duz' point to failings rooted in both technology and culture. They assert that the material difficulties attendant to economic and industrial backwardness retarded the development of the airplane and its attendant technologies; such material hindrances were in turn exacerbated by the unique failings of tsarist military culture. In this view, the atmosphere that prevailed at the level of policy making and leadership—again marked by inertia and apathy, attachment to routine, reverence for seniority, and petty intrigue—seriously inhibited exactly the types of institutional responses demanded by aviation: the necessity of foreswearing old technologies and established interests, reordering outlooks and priorities, seizing opportunities, fostering initiative and activity within the junior ranks of the officer corps, and throughout all a willingness not to be deterred by failure.[67] Emphasizing a favorite theme of Soviet historiography, Duz' asserts that the pernicious influence of Tsarist military culture was most strongly manifested in the army's deference to foreign aviation technology and a marked failure to develop indigenous designs, inventions, and industry. This obeisance to the West is seen as a function of the army leadership's short-sightedness, narrow-mindedness, and pervasive lack of confidence in its own people and resources, itself fostered by tsarist Russia's lingering backwardness.[68] This indictment of the senior levels of the army is fleshed out in popular histories and biographies of early aviation heroes that locate cultural resistance at the level of line as well.[69] These works emphasize such familiar iniquities as dilettantism, inertia, the narrow protection of parochial interests, red tape, biases based upon class and social background, and the pursuit of aviation as a fashion and fad. All of these failings, it is argued, worked to impede the committed activity of serious enthusiasts of military aviation.

Yet the themes of technology and community within the tsarist officer corps must also frame a reexamination of the Imperial air service. If the atmosphere within the military was not entirely propitious for the embrace of the airplane, this should not be altogether surprising. Most new technologies encounter initial resistance, as militaries tend to be inherently conservative institutions, and the airplane was no exception. In the context of the post-1905 period, it was particularly easy to view aviation as at best irrelevant to the army's pressing needs, at worst an expensive folly, and such attitudes were common among ranking officers. Yet even among the small group of enthusiasts who welcomed the building of a Russian air force, as well as the growing number of officers in the military at large and in the higher reaches of the War Ministry who would support their efforts, the arrival of the airplane recast the debate regarding modernization and technology, simultaneously pitching it at a higher level and expanding the scope and implications of its concerns.[70] Four factors stood out in this process. First was the undeniable symbolic significance of the airplane. This was no mere new gun or piece of equipment, but the culmination of humanity's age-old yearnings to conquer the skies. Observers quickly recognized that in terms of its power and promise, its potential and its demands, the airplane transcended the sphere of the military itself. Yet equally apparent was aviation's intimate relationship to the military. Indeed, two related assumptions regarding the airplane quickly became articles of faith amongst civilian and military proponents of the airplane, particularly in Russia: one, that its potential was enormous, its benefits assured, but also that its fullest realization was off in some near or more distant future; and two, that the most immediate manifestations and tangible applications of this potential were clearly military ones. Third, the incredibly swift and dramatic growth of aviation had a major influence in framing the responses to it. The rapidity of the technological development of the airplane, as well as the subsequent rapid growth of the services and personnel devoted to it in this period, daily transformed opinions and expectations. Lastly, the Russian army was not governed solely by its own assessments, but was constrained to respond to the increasingly rapid pace of military aviation development among the other great powers.

While drawing initially upon existing institutions and personnel, the army would thus have to develop new organizational structures, definitions of role and function, and standards of service for the needs of military aviation. This would challenge its ability to assimilate new technologies, institutions, cadres, and conceptions of duty. In turn, one can employ the paradigms of militarism and professionalism discussed above as a conceptual map to explore Russian officers' own emerging understandings of their role, function, and identity as members of the aviation service. To place these issues in human terms, one can trace the careers of two men who in many ways were emblematic of the experience of tsarist military aviation: Mikhail Efimov and Staff-Captain Petr Nesterov. In light of the enormous stature of Igor Sikorskii in the history of Russian aviation, due to his tremendous accomplishments in both the pre- and post-revolutionary periods, these choices might seem surprising. However, in the years before World War I, due to his civilian status as well as the independent sources of funding for his inventions, Sikorskii's work was largely self-directed and autonomous, despite his collaboration with military officials in testing and evaluating his designs. As such, his experience in many ways reveals more about his undeniable individual genius than it does the broader contours of military aviation policy.

In contrast, Efimov and Nesterov were intimately connected to the evolution of tsarist military aviation throughout its brief but significant lifespan.[71] The peasant son of a former army NCO and grandson of a freed serf, Efimov was a telegraph technician and avid sportsmen. Under the patronage of a leading Odessa banker and aeroclub enthusiast, in 1909 Efimov set off to be trained as a pilot in France, before there were any opportunities to do so in Russia. He quickly established himself as a leading pilot, extricated himself from the oppressive contract which bound him to his wealthy patron's service, and then pursued a successful and lucrative career as an exhibition pilot. Efimov quickly became the most popular Russian hero of the skies in this period, in part because he was one of the very first and very best, in part because unlike many he also managed to stay alive. Yet at the personal request of the Grand Duke Alexander Mikhailovich, Efimov abandoned his exhibition career to return to Russian and serve as an instructor training officer pilots at the new Sevastopol Aviation School.

Born in Nizhnyi Novgorod to provincial nobility and educated in a military cadet corps, Petr Nesterov was trained as an officer, finishing the Mikhailovskoe artillery school in 1906 as a lieutenant. Too late to see active service in the Russo-Japanese war, Nesterov nevertheless had ample time to mull over larger issues of defeat, reform, and modernization as he tended to his duties in an artillery unit stationed in the Far East. Seeking an opportunity to fashion a career of consequence and truly contribute to the rebuilding of the Empire's military power, he eventually gravitated toward the study of aeronautics and aerodynamics that had interested him since his childhood. Due to his initiative Nesterov eventually obtained a temporary posting to a Siberian aeronautical company. From there he would greet the news of the birth of the Russian Imperial air forces.

2| Flights of Fancy

In the summer of 1909, the Grand Duke Alexander Mikhailovich was vacationing in Europe when he had occasion to read of one of the signal events in the early history of aviation, the Frenchman Louis Bleriot's flight across the English Channel. Immensely impressed with the possibilities of heavier-than-air flight, the Grand Duke returned home to Russia and gathered the senior members of the Committee for Strengthening the Fleet Through Voluntary Contributions, a quasi-public institution he had founded to solicit donations for the reconstruction of the naval fleet which had been largely destroyed at Tsushima. The Grand Duke had seen the future, and it lay not on the sea but in the air. He convinced his colleagues that the funds already collected, as well as future donations, should go not toward the purchase of ships but of an aerial fleet.[1] The army, the autocracy, and the nation as a whole had seemingly been granted a new and powerful means to erase the lingering humiliation of 1905–06.

The Grand Duke was perceptive enough to recognize that the airplane, that quintessential symbol of the modern age, presented both unique challenges and opportunities to the tsarist army and to the Imperial state as a whole. It at once created new tactical and strategic considerations and even conceptions of geographical space; it would thus impose upon the army the need to develop an almost entirely new branch of arms in the midst of a pan-European arms race. The airplane was a technologically advanced weapon that would make sophisticated demands upon Russia's relatively underdeveloped industrial base; it thus would force the military leadership to consider how best to utilize the industrial potential and civilian resources of the Imperial state and society for the needs of aviation.

Lastly, as the response to Bleriot's flight graphically demonstrated, aviation had already ignited the curiosity and support of enthusiastic masses of citizens throughout Europe. While presented with this unique chance for redemption, Russia's from the very beginnings of aviation, Russia was integrally part of the broader experience of humanity's encounter with flight. Robin Higham notes that almost to a rule, in countries with very different social, economic, and cultural contexts and historical experiences, the birth of aviation was occasioned by strikingly similar responses: the handful of pioneers and dashing sportsmen, forums of aviation exhibitions and meetings, wildly enthusiastic publics, the projection of military power.[2] Throughout the world the airplane was seen as the apotheosis of modernity and the positivist view of the future, as proof of the ability of man to overcome nature through the force of his intellect. At the same time, the birth of the airplane, by virtue of the real and potential power it represented, was also perfectly digestible and applicable to the prevailing Darwinist world view of the competitive struggle of states and national cultures. As such, aviation signaled the beginnings, or perhaps the coming of age, of mass participatory politics and popular culture. While it may be a bit romantic to say that the airplane touched some common primeval chord in the collective soul, aviation and its apostles, the aviators, seized the popular imagination like nothing else. It is a hold which aviation, and its extension, space flight, has largely maintained to the present day.[3]

The Russian public, buoyed by a newly liberated popular press, would prove no less interested in the airplane, thus often forcing military aviation policy into a public spotlight. In his bold decision to abandon the sea, take to the air, and to use public funds to realize this dream, the Grand Duke announced himself ready to seize these opportunities and to meet these challenges. Enraptured with the airplane and bringing tidings of its future power, the Grand Duke did not return to a completely blank slate in his own land, however.[4] In the second half of the nineteenth century the Russian army had prepared various reports and commissions to study the question of aeronautics. The institutionalization of serious efforts to address the demands and military potential of aeronautics came with the founding of the Aeronautical Training Park in 1890 by a handful of enthusiasts of the engineering service, among them a junior officer named A. M. Kovan'ko. By the early 1900's the Russian army possessed several aeronautical units, consisting of various types of balloons. Some of these units, along with officers from the Aeronautical Training Park, took part in the Russo-Japanese War. Assessments of how well these units performed in Manchuria varied, but experimentation with balloons and dirigibles continued in the years after 1905, and the War Ministry continued to allot funds for training and purchases.

Then, in 1903, the Wright brothers flew, and by 1908 they were peddling their wares to the militaries of Western Europe. Other inventors and constructors, mostly French, soon provided competition, and the race to conquer the skies was on. In Russia, "although many senior military men saw airplanes as interesting if dangerous toys,"[5] the War Ministry lurched into action, prodded by advances made in Europe as well as the success of the First All-Russian Aviation Week, held in St. Petersburg in May 1910. In mid-1910 the Aeronautical Training Park was reorganized as the Officers' Aeronautics School [*Ofitserskaia Vozdukhoplavatel'naia Shkola*, hereafter OVS] with proposed aeronautics and general aviation divisions and a projected yearly output of 40 officer graduates.[6] At the same time the Aeronautical Section of the Main Engineering Administration, which was responsible for organizational and administrative aspects of military aviation and aeronautics, was expanded. Lastly, a special Aeronautical Committee was established to better address scientific and technical questions concerning equipment and design proposals.

In the field, aircraft of various types, including Wright and Farman, were purchased by the army and delivered to St. Petersburg in 1910. Two leading civilian pilots were invited to the Officers' Aeronautics School as instructors, whose first task would be to train the group of three OVS officers selected to be the permanent future cadre of the Temporary Aviation Section. Training flights with these officers, the first to be trained in Russia and not abroad, began in the spring of 1910. Finally, the Temporary Aviation Section of the Officers' Aeronautics School was formally established at the Gatchina aerodrome in the fall of 1910.[7] A second aerodrome, the Korpusnyi, built and equipped exclusively for military purposes, was established soon after in October 1910. Another sphere of War Ministry activity included providing financial subsidies to a few civilian Russian aeroclubs and aeronautical societies, most notably the Imperial All-Russian Aeroclub, founded in 1908 and based in St. Petersburg, and the Odessa Aeroclub. Such funds were used, with certain War Ministry stipulations, to hold exhibitions and competitions, train pilots, and propagandize. Money was also provided to a handful of aviation and aeronautics publications. The War Ministry also provided subsidies and advance payments to domestic inventors and designers whose projects were deemed to hold promise, as well as fielding requests for orders and protection from domestic commercial firms who wanted a piece of the military aviation pie.

Lastly, the organs of the War Ministry and the aeronautical service in particular also kept abreast of aviation developments abroad. Some information was gleaned from the Russian military attaches, who often received specific instructions and requests to report on aircraft exhibitions, maneuvers, or specific pieces of equipment. However, in a nod to the quickening pace of aviation development, a report of the Engineering Administration to the War Minister in Oct. 1910 noted instead the growing practice of sending observer and fact-finding missions abroad, particularly to France. Such missions included officer-engineers from purchasing and technical commissions, senior officers of the aeronautical service, and often professors and other civilians with relevant expertise in aeronautical science or industry. Such random missions were not enough, however, and the report recommended both a regular system of sending officers of the Aeronautics Department abroad to acquaint themselves with the latest technical advances, and that specialists accompany the senior command for attendance at maneuvers and reviews of foreign armies, as only in this way could they observe the technical units usually restricted to them.[8]

Overall, however, the early efforts of the army in the sphere of aviation can be described as somewhat tentative. Although the army leadership had considered making the Temporary Aviation Section a permanent division when it was established at Gatchina in 1910, this was not done until 1912. The Aviation Section actually welcomed its first class of trainees only on May 1, 1911, and this included all of seven officers.[9] War Minister Sukhomlinov had ordered the immediate formation of aviation squadrons in 1911, but only a handful of these planned squadrons were beginning to be formed. Official efforts to develop aviation continued to consist largely of experimentation conducted by the permanent staff of the Officers' Aeronautics School as well as the civilian inventors, constructors, and scientists who often worked closely with such officers and occasionally received military subsidies. For the most part, Russian military aviation remained restricted to the Gatchina aerodrome.

A somewhat more auspicious effort was put forth by the Grand Duke Alexander and the Department of the Air Force [*Otdel Vozdushnogo Flota*, hereafter OVF] a newly established division of the Committee for Strengthening the Fleet Through Voluntary Contributions. Convinced that "victory in a future war will be impossible without an aerial fleet,"[10] the Grand Duke threw himself into the task of purchasing an air fleet (as opposed to building one, in what was to become an important distinction) and training the cadres to man it. In 1909 the first group of eight Russian army and naval officers and seven sailors and soldiers were sent to France to train as pilots and mechanics, respectively. The Committee simultaneously arranged for the purchase of eleven aircraft of various types from France.[11] In the fall of 1910 the Committee formally established its own Aviation School at Sevastopol to train officers from all branches of the army and navy; the program then consisted of ten planes and 29 pilots.[12] Several of the school's officer-instructors went on to take part in the First All-Russian Festival of Aeronautics in September 1910, while squadrons from the school twice participated in army maneuvers over the next two years. Over the same period the Sevastopol Aviation School was to grow in size and move to a new and specially equipped aerodrome at Kach' on the outskirts of Sevastopol. The school also added lectures and theoretical courses in aviation and aeronautics to supplement the practical pilot training that was the expressed emphasis of the Department of the Air Force.[13]

There stood Russian aviation, in its general outlines, up until about the end of 1911. In truth the War Ministry had not done a great deal with aviation in its first few years. Whether it was negligent in this regard or appropriately cautious is debatable; it must be kept in mind that airplane technology itself was still quite new, usually unreliable, and quite often deadly. In contrast, the Grand Duke and his colleagues had worked diligently in the course of three years to realize his vision. However, more important than the concrete achievements attained in these years, or later assertions of opportunities missed, was the relationship between aviation and the social and cultural context in which it was received in the Russian army. Defensiveness, anxious self-examination, and condemnation of enemies real and perceived existed alongside urgent pleas for reform, recognition of opportunities to start anew, proud appeals to tradition, and confident prognoses of the army's ability to master future challenges. This contentious atmosphere exercised a profound influence upon the initial responses of Russian officers to the airplane. In many ways the tsarist military community would prove particularly receptive to the airplane and offer unique opportunities for its further progress; in other, telling ways this community and its culture would prove dismissive of aviation's promise and pose serious obstacles to its development. The airplane, in turn, would inspire both new hopes and novel fears within this community, while also eliciting new forms and channels of activity to meet the challenges aviation posed.

At the same time, however, the growth of aviation was again a profoundly international phenomenon. Russian officers would avidly keep abreast of developments in aviation amongst both allies and enemies, searching for models and practices to emulate or avoid. Their appraisals would draw upon the cosmopolitan experience of aviation discussed above while also revealing uniquely Russian conceptions of time, space, progress, and national identity. These were themes that would frame the subsequent development of Russian military aviation. It is thus not only instructive but absolutely necessary to first consider some of the initial responses to the challenge posed by the airplane among Russia's European neighbors. In France, institutional tensions within the military were accompanied by ever-increasing concerns about demographics, specifically the widening population gap between France and Germany. The intersection of these concerns became the focus upon the "offensive à la outrance" and the stress upon the role of morale in combat.[14] This institutional culture served to shape the French response to new technology. Ever conscious of the growing masses of German reserves, the French army was among the first to turn its attention to aviation. The gallantry and elegance of airplane flight seemed tailor-made to the romantic, voluntarist, offensive spirit of the idealized French officer the army sought to project. When gleefully compared to the huge, ponderous German zeppelins, it seemed like national and military stereotypes had come to life in the air. Moreover, in France military aviation became a striking intersection of the public arena and the military sphere. To the public at large, aviation's tangible heroes, the aviators, became powerful and palpable prophets in the gospel of revanche against Germany. Numerous civilian aviation organizations and clubs developed rapidly and played a leading part in channeling public efforts to aid military development of the airplane. Aviation thus became a powerful symbol of a nation in arms. John H. Morrow Jr. concurs with earlier commentators by noting that "this national passion for aviation was the essential context for the French army's adoption of the airplane."[15]

In Germany, the German General Staff's exploitation of the technology of the railroad to frame and perfect strategy is a legendary case study of the dynamic interplay of technology, institutional culture, and organizational adaptation. In general the German army seemed the most technologically and industrially oriented of the European armies. Yet in his analysis of the initial public enthusiasm over the zeppelin, Peter Fritzsche emphasizes that this was at first a phenomenon of the newly articulate middle classes, who proudly and self-consciously compared this enthusiasm to the moribund conservatism of the army, crown, and other powers that be that had initially ignored the efforts of Graf von Zeppelin. For this middle class public that donated millions of marks in national subscriptions, aeronautics and aviation stood as proud symbols of Germany's might and prowess in applied science and industry, as befitting the most economically and technologically dynamic European nation of the early twentieth century.[16]

Yet the German military, following an initial period of uncertainty and skepticism that is by no means uncommon when armies confront new technologies, eventually warmed to the task under public prodding. Although Fritzsche notes military utilization of the zeppelin was still cautious, hesitant, and unimaginative, and that most farsighted zeppelin enthusiasts continued to be civilians,[17] the Germans did better with and displayed more faith in airplanes. As Morrow describes, German aviation thus became increasingly militarized. The German army "effectively controlled civilian aviation organizations ventures through its pervasive influence in German society, playing upon chauvinistic notions of cultural supremacy to bolster military aviation," even to the point of manipulating public donations.[18] Germany was soon recognized as the second leading aviation power after France.

Regarding England, both Avner Offer and David Edgerton have offered variations of a compelling argument: that military philosophy and policy were guided by a model of what Edgerton defines as "liberal militarism." This entailed an embrace and focus upon technology, as opposed to manpower, and its most economical and efficient application in the waging of war. Offer structures his discussion within the context of British war planning before the First World War. Illustrating the navy's plans to cheaply and efficiently starve Germany to death or at least revolution, he makes the provocative argument that England, that most gentlemanly of nations, rated very highly in terms of militarism, construed in terms of its willingness to efficiently and ruthlessly employ technology and force.[19]

David Edgerton develops this argument in his own fashion with specific regard to the airplane.[20] He argues that the development and application of aviation fit perfectly within the emphasis and reliance upon technology, the search for material as opposed to personnel solutions for military challenges, and the view of war from a largely economic, cost-effectiveness outlook that was inherent in the "liberal militarism" of the English military establishment and society. Perhaps Edgerton's most intriguing observation is the seamless way in which a common commercial world view had penetrated even those segments of English society—the socially upper crust officer corps—deemed most conservative and hidebound. He bears this out in his discussion of the aviation services, illustrating the numerous points of contact between military, scientific, and commercial circles in the development of aviation. He thus argues that both the contemporary and historical opinion of England as the most "behind" of the great powers in the realm of aviation is simply untenable.

In Russia the situation, both in regards to a generalized institutional interest or approach to war and a prevailing hegemonic outlook such as commerce or revanche, was rather more ambiguous. Instead of the burning desire for revenge of the French, the confident arrogance of the Germans, or the languid self-possession of the English, the Russian army was rent by serious division and debate regarding its needs and priorities, its military role and function, and its very raison d'etre within the structure of the Empire. As noted, various attempts to define and cultivate a fixed military doctrine, a unifying ethos, or a coherent set of institutional interests came to naught, given the divisive social and cultural context of the army as well as the resistance of officers and members of the dynasty. Such disputes over the terms and limits of modernization were again manifested sharply in regard to the issue of technology. While it was difficult if not impossible to deny the importance of technology in modern warfare, there was wide latitude to interpret the scale and scope of that significance. Those who advocated a more vigorous exploitation of the technological sphere were opposed by others who emphasized traditional military virtues and the role of morale. Each side invoked their own standards and conceptions of progress and backwardness, adaptation and contamination, warning of dire consequences if the wrong choices were made.[21]

For believers like the Grand Duke, the airplane promised a veritable revolution in warfare. There were progressively numerous expressions of this faith. In the War Ministry's Annual Report for 1908, the Main Engineering Administration concluded that although airplanes "can at present still not make very long flights or rise to any great heights, and while in general they are not suitable for military purposes, in the future they will nevertheless play a tremendous role in military affairs and so will undoubtedly be introduced into the armament of the army."[22] In a November 1908 report to Colonel Kovan'ko, Captain German, a senior officer of the Aeronautical Training Park and editor/publisher of the journal *Vozdukhoplavatel'* [*The Aeronaut*], began by noting that recently aeronautics had acquired a "very important significance" and that very recently airplanes have achieved such a level of perfection "that the question of their application for military goals is a question of the near future."[23]

Two years later, Colonel Kovan'ko himself echoed this opinion forcefully in an interview for a newspaper article entitled "The Airplane in Future War". In it he provided an appraisal of current airplanes and equipment, noting that "the progress of aviation is rapid in the extreme... The time is not far off, when aviation will become a fearsome weapon."[24] Soon after the War Ministry journal *Voennyi sbornik* printed a piece devoted to coverage of the 2nd International Aeronautical Exhibition held in Paris in October 1910, at which it was noted the Grand Duke Alexander made several appearances.[25] Describing the numerous wares and devices displayed there, the piece concluded that "the military significance of all of the apparatuses described is clearly obvious" if only from a cursory look at the excellent photographs of the ground taken during flight. Such convictions were summed up two months later by V. Bronskii, regular commentator on sport and aeronautics for *Voennyi sbornik*: "the conquest of the air moves forward with rapid steps and the application of flying machines in military affairs on the very broadest scale is now becoming a question of the very nearest future."[26]

As these quotes indicate, there was an early if hazy recognition of the potentially awesome destructive power of the airplane, even if practical applications were admitted to be off in the future. As elsewhere in Europe, civilian organizations, unlike officers, were under no pressure to be held accountable for their predictions nor to produce results in this regard. As such, free of the need to demonstrate restraint, they provided the more bellicose utterances. Thus, Von Hardesty cites various articles that appeared in 1910 and 1911 in the newspaper *Moskovskie vedomosti* with titles like "War and the Air Fleer," "Bombing from the Air," and "The Military Significance of Aeronautics." These articles warned, among other things, of "murderous projectiles" dropped from the sky.[27] In 1912 the Imperial All-Russian Aeroclub, whose members included a good number of senior officers as well as enthusiasts from the aeronautical service, would publish the propagandistic pamphlet *Aeronautics*. The pamphlet reviewed how the Italians had been the first to employ airplanes in both reconnaissance and in bombing enemy troops while fighting the Arabs and Turks; it observed that "the linen bird became the carrier of death" and concluded that "already by these results it is possible to judge the truly important significance of military aeronautics."[28] In less poetic language, *Razvedchik* would later concur by noting that "the combat application of airplanes in the past Italo-Turkish war demonstrated their complete suitability as apparatuses serving to drop bombs."[29] In all of this aviation offered, instead of the chimera of the force of morale, the tangible premonition of the exercise of power on a grand scale.

Admittedly, however, this lay in the future. As in the other European militaries, the airplane was initially envisaged in Russia primarily as a vehicle for reconnaissance and communications. Here too, however, the airplane appealed to uniquely Russian concerns and needs. The appraisals of the airplane's potential were magnified in the context of the relatively poor transport and communications infrastructure and the broad and diverse expanse of the Empire. In his summary report of 1911, the commander of the Turkestan military district appealed for the establishment of an aviation school in Tashkent on such grounds, arguing that

> the exceptionally advantageous climactic conditions of the Turkestan *krai* and the mountainous character of the terrain, which makes reconnaissance difficult, in all possible theaters of our military operations in Central Asia forced me in my last summary report to express my desire for the organization of training military pilots on airplanes in Turkestan.[30]

Aviation thus appeared to offer the military the means to get around seemingly intractable difficulties of geography, limited financial resources, and Russia's own late start in the area of railroads and the telegraph. Moreover, in the new and unspoiled sphere of the air, the accumulated legacies of backwardness on the ground need not prove a hindrance to future progress.

It was not only in its novelty and fearsome potential that aviation was unique in the broader debate regarding technology. If the airplane seemed destined to revolutionize the practice and conduct of war, it promised to be equally transformative for officers themselves. While the airplane itself held a material and symbolic significance as the centerpiece around which the aviation services would be constructed, it did so in a way which served to reinforce the Russian emphasis upon human resources. Even with this most modern of machines, as much or more attention was given as to who would fly them; nothing was so useless as a plane without a pilot. A March 3, 1910 report of the Main Engineering Administration to the War Minister summed up the situation thus:

> the rapid progress in the development of flight in the air and the possibility of applying it in military affairs has forced all states to turn the most serious attention to this subject. Experience has shown that the success of flying in airplanes to a significant degree depends on the experience of the aviator, regardless of the type of flying apparatus...[31]

Thus, emphasis was given to creating a cadre of technicians, mechanics, and above all pilots to staff the aviation services. This entailed attempts to identify and cultivate knowledge, skill, and expertise in the theory and practice of aviation. At the same time, the union of man and machine in the relationship of pilot and aircraft seemingly rendered irrelevant the opposition of passivity and voluntarism, morale versus technology, which marked debates between artillery and the bayonet. Moreover, it was officers themselves who were to be the dominant actors in this relationship, somewhat nullifying the concerns about intelligence, passivity, and motivation that implied much of the debates about the relationship between soldiers and new technology.[32]

Instead, initiative and voluntarism were to be the dominant themes of officers' engagement with the airplane. These were qualities that had distinguished the clubby atmosphere and small officer cadre of the Officers' Aeronautics School ever since its inception decades earlier as the Aeronautical Training Park. Unlike advances in artillery, aviation was almost entirely new, while both the training and experience of those involved in dirigibles and balloons was only partially translatable to the sphere of heavier-than-air craft. Nevertheless, with its tradition and credentials well established, the staff of the Officers' Aeronautics School was deferred to as the natural constituency to appraise the airplane. The deference and status accorded to OVS officers drew upon Russian military culture more broadly. While notably below the Guards and cavalry branches in the hierarchy of military prestige, the artillery and engineering services (of which OVS officers were a part) nevertheless commanded a certain amount of respect for the specialized knowledge and training they possessed. Such attitudes dovetailed with broader positive valuations within Russian culture and society of scientists and technicians, as opposed to those involved in commerce and business.[33] This appreciation for the specialized knowledge and expertise of such officers went hand in hand with a certain amount of deference and a willingness to allow them a good deal of autonomy within their own specialized sphere.

Thus, in one of the very first official responses to the birth of aviation, in 1908 the General-Inspector of Artillery, the Grand Duke Peter Nikolaevich, ordered the officers of the Aeronautical Training Park to construct an airplane themselves in lieu of purchasing machines and instruction from the Wright brothers. This decision was made in response to a report from Col. Kovan'ko that the money would be better spent thus.[34] As the officers in question had never flown in heavier-than-air craft, nor ever even seen an airplane, perhaps it is not surprising that the machines they subsequently constructed—five in all— failed to fly. Yet this was symptomatic of a deference to experts and a reliance, even if overdone, on their expertise. Despite this failure, many officers of the Officers' Aeronautics School continued to individually design and test various inventions. Colonel Ulianin, a balloonist at the OVS and later the first commander of the Aviation Section, was active in glider construction and experimentation; Colonel Kovan'ko's son, who trained at and was later assigned to the staff of the OVS, likewise constructed an experimental aircraft.[35] Moreover, several officers of the both the OVS and the Aeronautical Section of the Engineering Administration were editors or regular contributors to aeronautical periodicals.[36]

Thus, the airplane initially appealed to established constituencies and recognized experts within the officer community, and it was to these that the army first turned. An alternative model of engagement with aviation was presented by the activity of the Grand Duke Alexander and the Department of the Air Force. In pointed contrast to the ill-fated experiments at the Aeronautical Training Park, the Grand Duke decided immediately to defer to the expertise and experience of foreigners, choosing to both purchase aircraft and initially train officers abroad. These officers in turn were selected from all branches of service, not just the engineering services. In the Grand Duke's great crusade of aviation, enthusiasm and commitment were valued more than seemingly out of date knowledge and specialized training.

The contending philosophies which underlay these differing responses to the airplane were difficult to mask and would continue to make themselves felt. Aviation thus offered a forum for activity and commentary to all, not just the officers of the engineering and aeronautical services. While there were of course the naysayers who mocked the airplane, there were no shortage of officers who were willing to persuade or be persuaded of the benefits of this new technology. Thus, engineering officers were joined by numerous line officers in the various public aviation, aeronautical, and technical organizations. Of course, some of these officers were merely honorary or passive members, but the number who were actively engaged in the work of such organizations, especially at the level of colonels and captains, was impressive.[37] Perhaps of greater consequence were senior officers outside the aeronautical services who were convinced by their experience, opinion, or ambition of the power and importance of aviation. The best example of this phenomenon was the activity of the Department of the Air Force and its leaders, the Grand Duke Alexander Mikhailovich and his assistant General Kaul'bars, commander of the Odessa military district. Kaul'bars, despite holding no official post in the aeronautical service or the engineering corps, was instrumental in spreading the gospel of aviation through both his chairmanship of the Department of the Air Force and the provincial aeroclubs he helped found.

Even within the sphere of the army proper, initiative in the development of military aviation was often exercised not by the central organs of the War Ministry but locally and individually. Notable in this regard were several military district commanders in addition to Kaul'bars. Often defining military necessity in their own terms, they took steps to see to it that such needs were met. The doctrinal freedom which allowed them to do so was aided as well by the specifics of tsarist military administrative structure, where military district commanders, particularly in the border districts, had a good deal of discretionary authority in assessing needs and priorities, even in war planning.

Already noted, for example, was the letter of the commander of the Turkestan military district urging the need for a military aviation school in Tashkent; he went on to set up his own aeroclub and school and subsequently requested War Ministry subsidies.[38] In a 1910 letter to the War Minister, Gen. P.A. Pleve, the commander of the Moscow military district who with the War Minister's permission also doubled as the chairman of the Moscow Society of Aeronautics, also requested subsidies for a new special Military Committee of the Society. Pleve emphasized that the task of this Committee's activity would be "to serve the needs of military aeronautics, as a means of resolving various theoretical questions connected with the application of aeronautical devices in the sphere of strategy and tactics and primarily through the practical study of military aeronautics, carrying out flights and preparing a cadre of experienced military aeronauts." In his opinion the Committee's activity could "doubtless be of great use to our army, in particular keeping in mind the absence of military aeronautical squadrons in Moscow, the geographical situation of which—in the center of a vast continent—is extremely advantageous for carrying out flights.'[39]

There was a last group to which the airplane beckoned: the long-suffering officers of the line, traditionally seen as the exemplars of a culture of apathy, ignorance, and small-mindedness. At a time when few aspects of an average officer's life, from public perception to material circumstances, could be considered heroic, aviation appeared as a fabulous dream, a chance to achieve distinction, fame, or even, at a more prosaic level, escape. Open to officers from all branches of arms, the Temporary Aviation Section of the Officers' Aeronautics School and the Sevastopol Aviation School offered deliverance to those willing to pursue it; the airplane literally offered a chance to rise above the monotony of regimental life. A humorous piece from a December, 1910 issue of *Razvedchik* told the fictional story of a regimental chancellery officer who one day observed a colleague ("a very well-mannered second lieutenant Sladkov"), who, while engrossed in accounts concerning among other things wheel grease, suddenly exclaimed in a sobbing shout "'no, I can't anymore!'" and got up and left, headed in an unknown direction. The regimental colonel, carefully counting the days until his retirement, barely looked up.

After spending the night dealing with more of the minutiae and vagaries of regimental paperwork and recognizing the accounting mistakes he has made, the chancellery officer realizes he too needs an escape, and his conclusion is aviation. He thus composes a report to the command:

> 'to the battery commander. I report that I wish to study aviation and add that I agree to serve in the air not only for the required two years, but for all my life until retirement or until such time as, on legal grounds, I fall on my face. In fulfillment of this I ask: to acknowledge my vision sharp, my hearing keen, my heart strong, and I myself recommend me as outstanding. With <thermal engines> I am hardly familiar, but I feel their influence on me, as I have sat in the stockade three times. But when *correspondence* is introduced in the sky, and if, in the process, *aero-chancelleries* with staffs and *aero-stockades* are developed—then I will ask to dismiss me back to the earth, as all things being equal, it will nevertheless be calmer on the ground...'

After submitting his report, the officer discovers that his colleague, Second Lieut. Sladkov, had already flown as a passenger at the aerodrome the previous day. He concludes that while it is now the regimental colonels' turn to take to the air, "he probably will endure his year and a half."[40]

More romantically, the airplane offered the opportunity to achieve lasting glory. Emblematic of such opportunities were the huge and wildly enthusiastic crowds who turned out at the First All-Russian Festival of Aeronautics, held in St. Petersburg in September 1910. Organized by the Imperial All-Russian Aeroclub with the aid of a 25,000 ruble subsidy from the War Ministry, the festival was intended to rival the mass public aviation meets common in Western Europe. As noted earlier, several of the officer-instructors of the newly established Sevastopol Aviation School participated in this event, together with leading Russian civilian pilots. The daring of the pilots and the ecstatic reception given them by thousands of spectators seemed to assure a future of fame and renown for the new cadre of military aviators.[41]

While the Festival provided the opportunity to reach new heights of awe, it offered an occasion to plumb the depths of despair as well. This came with the crash and death of Lieut. Matsievich, one of the original group of Russian pilots trained in France, who on the previous day had enraptured the crowds with his brave and skillful flying. As noted in the introduction, Scott Palmer devotes a fascinating and compelling article to the death of Matsievich and the huge state funeral which followed. He asserts that the massive outpouring of sympathy for Matsievich gave the public and especially the press the opportunity to assert national identity.[42] Yet, while Palmer's analysis is fascinating, I think he does not stress enough the fact that this pageantry, this mass catharsis, was for a *junior officer*. When we consider the common assumptions regarding the negative attitudes to the military among broad segments of Russian society, attitudes largely returned in kind by officers themselves, this is particularly striking.

The promise of such glory in the air, even if the ultimate price was death, was trumpeted to all officers in an article in *Voennyi sbornik* enthusiastically entitled "Forward!" In the piece, K. Bronskii asked his readers to "look, look, at the terrible dark figure, clinging to the back of the courageous pilot, at the horrible smile of the specter of death, who extends the branch of victory and glory, which draws the fearless person forward and forward into the until now unreachable heights of the boundless aerial ocean." Whether the brave were able to wrest victory from the "bony hand" of death or perished in the attempt did not matter: "where one dies, two new fighting men will stand to replace him; where two or three lay down their lives, four, six new ones will take their place..."[43] The fate of Capt. Matsievich seemed to prove that such sacrifices would not go unnoticed or unappreciated by the army, the state, and the nation as a whole.

The funeral and commentary regarding the death of Matsievich highlights another significant point: this was the recognition early on, not least amongst officers themselves, that aviation was to be a very public technology. In terms of potential and promise, resources and effort, aviation clearly transcended the immediate sphere of the army itself. All of the officers quoted above couched their commentary in the broader context of the relationship of military aviation to the Russian state and public. The Main Engineering Administration report to the War Minister of March 3, 1910, cited earlier was typical. While it stressed the importance of military aviation, it continued in the same breath that "therefore in all states, besides the training of military pilots, serious attention has been paid to the development of the sport of aviation and to the preparation of as large as possible a number of persons able to pilot flying devices with the goal of using them in the event of war."[44]

At times the military directly addressed the public regarding the latter's assistance in the affair, as in the order issued by the commandant of the Sevastopol military garrison in November 1910. Announcing the birth of aviation and aeronautics, the commandant provided the soldiers and civilians of Sevastopol both edification and patriotic exhortation.[45] He noted that at present in Russia, as in all other foreign states, "aerial flying machines" were being introduced for the needs of the military; thus, as "this affair is new, difficult, and dangerous, and as it has enormous significance for the military power of the State, all residents should render help and assistance to military aeronauts when they need it." To aid such efforts the order first described the three types of flying machines—airplanes were described as having "the appearance of a large bird with outstretched wings and a tail"—and then gave details and rules for notifying the authorities of flights and assisting aeronauts if necessary.

Colonel, later General, Kovan'ko, commander of the Officers' Aeronautics School, envisaged a far more active role for civilians than assisting downed aviators or even being used as trained pilots. He was a particularly eloquent and persistent advocate of the view that in terms of its novelty, rapid development, technological sophistication, and industrial demands, aviation by necessity required the mutual and harmonious efforts of military, state, and civilian figures and institutions in order to reach its fullest, and especially fullest military, potential.[46] In the interview he granted to *Birzhevie vedomosti*, he expressed his certainty that Russia, having already produced excellent pilots in a short time, would turn out as well a series of talented inventors of military technology. Kovan'ko thus stressed the importance of developing a domestic Russian aircraft industry, both in terms of materials, design, and labor.[47] As if to underscore his views, the article in question also included interviews with Mikhail Efimov, again one of the foremost civilian Russian aviators at the time, and A. I. Guchkov, chairman of the Duma's Defense Committee and a man avidly interested in maintaining the military power of the Empire.

Pleas for public interest and assistance were echoed by the Department of the Air Force, which in its appeals for contributions self-consciously appealed to an idealized and united public opinion—that Russia should be great. As reported in *Sevastopol Illustrated Aviation Journal*, "the Committee hopes that all to whom the military power of Russia is dear will bring their kopecks and rubles to the cause of creating an air force, which in the future, in war time, stands to play an important role."[48] Such views were later phrased rather more eloquently by the Imperial All-Russian Aeroclub [IVAK] once it too had received permission to collect contributions. In its pamphlet *Aeronautics* it presented aviation as a new and unique opportunity to bypass Russia's unique problems of geographic expanse and poor transportation and communication infrastructure, as well as a means to augment military might. In its "Appeal" it proclaimed that "the IVAK hopes, that all to whom the honor, glory, and greatness of Russia is dear, who value its cultural prosperity and military might, will not stop before material sacrifices to the great cause of the creation of an All-Russian air force..."[49]

The civilian public was not to play a merely passive role. To begin with, the line between civilian and military activists in the aviation services was fluid and interaction constant and frequent. As noted, the leading aviation clubs and organizations, particularly the All-Russian Aeroclub, were heavily represented by both high civilian officials and bureaucrats and senior and junior officers of the aeronautical arms and regular army. Ties were also quite close with the handful of leading Russian civilian engineers, inventors, and builders involved in aviation—Sikorskii, Gakkel, Slesarev to name a few—who regularly consulted with military officers, utilized military facilities for their projects and experiments, and eventually sold their wares to the War Ministry. Military aviation was further aided by the energetic activity and experimentation conducted at civilian institutions and laboratories throughout the Empire by noted academics such as N. E. Zhukovskii, later dubbed "the father of Russian aviation" by the Soviets, S. A. Chaplygin, and K. E. Tsiolkovskii.[50] The patronage of the prominent played a part as well. The Kuchinskii Aerodynamics Institute was established in 1904 through an endowment by the leading industrialist D. P. Riabushinskii, while on the basis of a large donation by a French subject, Vasilii Zakharov, the Department of the Air Force was able to organize theoretical courses in aviation and aeronautics for its officers at the St. Petersburg Polytechnic Institute starting in 1911.[51] These men and institutions embodied the exercise of civil initiative and its contribution to the needs of military aviation.

Lastly, from the beginning of military aviation its officers were a regular feature in the popular, and to a lesser extent, the military press. There was enthusiastic regular press coverage of air shows and exhibitions, manifested most vividly in the burst of commentary regarding the death of Matsievich, travelogue type visits by civilian and military correspondents to the aviation schools, and the creation of numerous aviation enthusiast and technically oriented magazines. *Sevastopol Illustrated Aviation Journal* reflected the public mood when it announced "we can with pride announce, that at the present time we have military aviators who are not in the least inferior, in all respects, to the aviators of Western Europe.... Aviation is the only sphere of activity in which we do not at all lag behind the European powers."[52] Officers for their part reciprocated. The interview granted by Kovan'ko to *Birzhevye Vedemosti* cited above was but one example. Leading figures such as Colonel Ulianin of the OVS and later the famous Petr Nesterov granted interviews, explained their latest exploits, and participated in exhibitions designed to educate the public about the capabilities and virtues of the airplane.

At first glance optimism was the prevailing feature of this cooperation between army and public, both among the officers and among those in the civilian sector championing and publicizing their efforts. The Imperial state, to a greater extent civil aviation organizations, and officers themselves touted Russian aviation and what they considered its lofty position as proof positive that Russia was indeed a member, and an influential one at that, of the family of modern, "civilized" nations. This was a theme repeatedly expressed in the aviation journals and propagandistic pamphlets, as exemplified in the appeals of the Department of the Air Force and the All-Russian Aeroclub cited above. Thus, at a time when it seemed that the gulf of mutual understanding and respect separating the military and the public had never been wider, aviation and the nation's enthusiastic embrace of it seemed to offer a path to that elusive unity of patriotism, the perceived lack of which so often surfaced as a lament among Russian officers. Technology in tune with the best national characteristics—intellectual, moral, spiritual, martial—was the dominant theme; it was framed in an idealized definition what it meant to be Russian, usually opposed to an equally archetypal conception of the West. Thus, it was not just a matter of national power but of a national people. Robin Higham observes that the earliest heroes of aviation were most often represented as, and perhaps selected because of, the embodiment of idealized national values and cultural stereotypes. The best example was the United States' Wright brothers: self-made, independent, ruggedly individualistic American pioneers on the wild frontier of aviation. This was at least the popular perception, if not the reality.[53]

The Russian ideal of both national virtue and civil-military cooperation was embodied in the person of Mikhail Efimov. As noted, Efimov was already wildly popular among the Russian public as a native hero of the skies whose skill and bravery were second to none, yet whose virtue and nobility of spirit the shallow Europeans could not match. Many aviation officers shared fully in this broader conception. Colonel Kovan'ko appraised Efimov thus:

> the magnificent Efimov is an ideal flier, a gentleman in all respects... It would be a shame to lose him... How he does not resemble those foreign racing drivers who fly

> only for the sake of money... I have had to go abroad for all kinds of holidays and competitions... To be perfectly honest I will say, that Russians fly no worse. They are distinguished by the enormous presence of spirit, bravery, love of their affair, and the complete absence of monetary considerations.[54]

In 1910, at the personal request of Grand Duke Alexander Mikhailovich, Efimov selflessly and patriotically abandoned his lucrative performance career in Europe to return home and take a post as chief instructor at the Sevastopol Aviation School. He had indeed been anxiously awaiting an opportunity to serve the cause of aviation in his own country.[55] From contemporary accounts and subsequent historiography, it is clear that Efimov could represent different things to different people. Soviet historians and memoirists who later joined the Red Army stressed his proletarian origins and the self-made nature of his achievements. The contemporary press, across the social and political spectrum, proudly emphasized the fact that he had demonstrated Russia's superiority in the air to his former French masters. But what united both of these views, and what gives suggestive hints to the unifying force of an idealized nationalism in Russia during this period, was the celebration of Efimov as a hero of the Russian people, a man who asserted Russia's pride, place, and power before the often skeptical European powers. He thus could be claimed by all Russians on that basis. Of course, the implied vision of an idealized Russian nation varied greatly among such papers and publicists, but the focus was the same—Efimov.

As symbolized by Efimov, there was seemingly no shortage of patriotic, altruistic Russian subjects who would willingly aid the army in the great task of maintaining the place of the Empire as a major cultural and military power. Provided with such assistance, there seemed to be few limits to what Russian military aviation could achieve. The greater ramifications were equally clear. A. Poltaratskii observed in *Razvedchik* that soon "*an age of steam and electricity* will, finally, be replaced by *an age of flying people*, *bird-people*, the dream of humanity, as ancient as itself, will be fulfilled, and those horizons will open, of which the most fervent imagination of novelists did not dream."[56]

Nevertheless, despite the constant experimentation and technical advancements, the bold self-assurance of the new aviation officers, and the confident prediction of a union of the efforts of army and people, supporters of Russian military aviation still faced unique and serious challenges. Competing strategic priorities, scarce resources, institutional inertia, the prerogatives of established branches of arms, and the frailty of the technology itself all combined to produce continuing resistance or studied apathy regarding the airplane. More broadly, the particular conditions of the Empire—its underdeveloped economy and industrial base, as well as its relatively low level of technological and infrastructural development—presented special impediments to the development of both military and civil aviation. And while aviation promised the army new, almost unlimited opportunities, it created an entirely new set of worries and problems. The nature of these concerns largely mirrored those of the airplane's perceived benefits, again illustrating universal apprehensions attendant to the birth of the airplane as well as fears and anxieties unique to the Russian military community.

In a fundamental sense, the power symbolized by the airplane served to reinforce fears of the dangers of backwardness. Chief among such concerns was dealing with the air forces of neighboring, enemy nations. The traditional emphasis upon defense in depth and the ability to trade time and territory—itself under attack by some Russian military theorists for the passiveness and perhaps weakness such a strategic inclination represented—was, if only partially recognized at the time, being violated by the airplane. For if aviation promised to more securely bind the Empire together, it also promised to bring Russia that much closer to its enemies. Throughout 1911 *Razvedchik* warned of potential foreign espionage conducted in balloons and airplanes and published official instructions regarding measures to be taken, including the shooting and capture of foreign machines and personnel.[57] The psychic shock was perhaps not quite as profound as in Great Britain, which saw the treasured isolation and security provided by its island status suddenly violated by the airplane, but the anxieties were similar in nature.[58]

Moreover, when propagandists waxed poetic about the potential destructive power of a Russian air force, it was not too difficult to invert this vision and imagine the view from the ground. Indeed, dystopia, apocalyptic visions of the destructive power of the flying machine were part of the dream of aviation almost from its birth, and in this Russia was no different. The opportunity to bomb cities, for example, was foreseen early on. The articles in the Russian popular press which warned of "murderous projectiles" dropped from the air could easily envision St. Petersburg or Moscow as their target. Such apprehensions were magnified within the context of the ambivalent resignation regarding the inevitable outbreak of a European war, for which many tsarist officers were by no means certain that Russia would be prepared.

This current of insecurity was extended as well to the proposed partnership of nation and army. With the traumatic memory of the defeat at the hands of the Japanese still fresh, the pleas from both officers and civilian organizations to advance aviation were tinged with predictions of potentially dire consequences if the benefits of the airplane were not actively pursued. Underlying these pleas was an implicit recognition that Russia's place in this family was tenuous, that it was not quite so civilized, and a fear of falling behind, perhaps permanently being left behind and sense forfeiting its place among the "civilized" family of nations. Thus, aviation would be not just proof of the achievement of "modernity" in Russia but a means to achieve a more secure and unassailable level of it. The pamphlet *Aeronautics* of the Imperial All-Russian Aeroclub was emblematic of this perspective. While praising the benefits of the airplane, it noted that

> given such rapid growth of the significance of the air force which we now observe, for each of the states it is a task of paramount importance not to fall behind the others in the matter of the organization of air power. And more so than for any other country this should be urgent for Russia. The blow inflicted upon our national pride by the last war places upon us the duty to work diligently on the recreation of the greatness of Russia as a strong military power...[59]

Colonel Kovan'ko of the Officers' Aeronautics School also repeatedly stressed the dangers of haphazard efforts, lack of common goals, and time wasted in such an important military and state matter as aviation, while the Grand Duke and the Department of the Air Force consistently warned of the dangers of falling behind in the air.

Yet perversely, the transformative potential of the airplane was limited by the very backwardness it promised to overcome. An immediate issue was how to produce or procure the aircraft and related technology to meet the perceived needs of the army, as it was quickly recognized that the existing economic and industrial resources of the Empire were poorly suited to meet the technological demands presented by the airplane. This was a situation that was becoming increasingly problematic for the Russian army generally in regard to advanced armaments; in the case of the airplane it presented itself in particularly acute form. Thus, a split appeared early on in response to this particular dilemma of "backwardness." There were those who argued that relying primarily upon products of foreign manufacture represented the best return on scarce capital; others argued that developing a strong domestic aircraft industry represented the only long-term, truly decisive solution. The first camp was prominently represented by the Grand Duke and the Department of the Air Force; the second by General Kovan'ko, the Officers' Aeronautics School, and domestic entrepreneurs and inventors. This debate over the desirability and plausibility of relying upon foreign or domestic resources was thus to prove one of the enduring and contentious throughout the short history of Imperial Russian military aviation, and will be explored at greater length in the next chapter.

This debate further betrayed an ambivalence regarding the vision of harmonious cooperation between the military and the public. As with the potential benefits, the potential dangers of aviation were perceived in terms of not only material but human resources as well. If officers by and large recognized the value and necessity of civilian involvement and effort, many harbored doubts about civilians' willingness and ability to render this aid on the army's terms. Distrust of outsiders, latent fears of internal and external traitors, and suspicions regarding the patriotism and loyalty of various groups and classes were all manifestations of pervasive anxieties among tsarist elites in the last decade of the Empire. The officer corps, which had been in the front lines of war, revolution, and repression in 1904–06, experienced these anxieties particularly acutely. Such conflicting attitudes and desires were to be a constant source of tension both among various strata of the officer corps and between the army and the civilian aviation institutions.

Thus, there was notable apprehension regarding the potential violations of the unity and integrity of the officer corps represented by the airplane. Such threats could come from both within and without. Pressing fears coalesced around the idea of the airplane falling into the wrong hands, in which it could promote sedition and internal threat. The non-Russian nationalities were obvious suspects in this regard. To forestall any dangers associated with the airplane, restrictions upon service in various geographic areas or sensitive technical units for such non-Russian groups were supposed to be extended to the aviation services as well.[60] This was particularly true for the Poles, who, judging from memoirs and anecdotal traces in official documents, were even less popular among the officer corps than they were in the state apparatus. Memories of 1831, 1863, and 1905, where Polish officers were perceived to have distinguished themselves by revolutionary and rebellious activity, died hard. This was but one aspect of broader fears regarding the loyalty of the Polish people, among others, at large. Such fears were manifested in the decision of the central military command, overriding the Warsaw military district command, to express their opposition to the proposed establishment of an aviation school in Warsaw under the auspices of the private aviation society "Aviata," even though the society had worked out its plans with the military district staff and proposed to train military aviators as well. A November 1911 report of the General Staff to the War Minister regarding this matter concluded that "the creation of private aviation schools in the western border region, where the native population in the majority of cases adheres to an oppositional way of thinking in regard to Russian statehood, is undesirable, as such schools and societies can serve more harm than good to the War Department."[61] Apparently, the specter of trained Poles flying in the skies of the Empire was not a reassuring one.

In another sense, the new aviation service promised both to reinforce and to disrupt established structures and paradigms of service and identity. Despite increasing attempts by the army command to encourage officers to integrate better the use of arms in this period, especially the coordinated use of artillery and infantry, it seems the prevailing view continued to be that technology was best left to the experts. This notably can be seen in the position and status of the Officers' Aeronautics School. Yet Peter Kenez observes that conditions of service in the small railroad and engineering corps, with only about 2,000 officers, were so different from the rest of the army that they led a very separate existence; he notes an observation made by Gen. Denikin in *Staraia armiia* that such officers "did not feel they belonged to the officer corps, and the other officers also regarded them as not really military men."[62] The staff of the Officers' Aeronautics School, trained engineers and aeronauts first, would no doubt have vehemently objected to such a description as they diligently worked to master the skies, as would their fellow officers from other branches of arms who had occasion to observe their pioneering efforts. Yet it was true that such officers had long held a virtual monopoly on expertise and activity in the field of aeronautics, a status that allowed them a certain freedom from outside scrutiny and interference and a marked deference to their special knowledge, skills, and talents. Emblematic of this status again was the early unsuccessful attempt by such officers to build their own airplanes rather than purchase them abroad. At the same time, this special status likely promised a limited allocation of resources, interest, and attention from outside the aeronautical services in the first years of the airplane's existence.[63] Moreover, such status could produce studied resistance to alternative paradigms which perhaps threatened established interests and constituencies. Notably, the active and enthusiastic Petr Nesterov saw several projected designs for aircraft and related equipment he submitted denied by the Engineering Administration in 1911.[64] Still serving in an aeronautical battalion in the east, his efforts to transfer to aviation training were also thwarted.[65]

Yet the Officers' Aeronautics School did encounter a competing arena of aviation activity in the Department of the Air Force and its royal patron. Imbued with their own sense of mission and urgency, free of the shackles of routine and institutional memory, and governed by their own conceptions of priorities and necessities, the Grand Duke and his colleagues vigorously set about building a powerful air force and training a new cadre of military aviators. The principles which guided their activities were fundamentally quite different from the established customs of the Officers' Aeronautics School, and for that very reason would draw criticism both from military circles and the civilian public. Dissension came from within the ranks of the school itself as well. For example, it is noteworthy that despite his undeniable talents and sense of patriotism, Mikhail Efimov, instructor at the Sevastopol

Aviation School, was not particularly well-received by all aviation officers, both on the grounds of being a civilian and being a plebeian, and because he was in a position of superiority over them. Moreover, he served at a school which was itself the product of the Grand Duke's vision and his semi-public organization's funds, and thus in part served as an indictment of the army's own lesser efforts in the field of aviation.

In its formative years the airplane as a technology and a potential weapon was met with both positive and negative reviews within the tsarist army, but neither was dominant. Such mixed appraisals of the military significance of aviation were common throughout the armies of Europe.[66] However, the unsettled institutional and cultural context of the tsarist officer community provided a unique cast to this phenomenon in Russia. As exemplified by the response of the Grand Duke, the bitter legacy of defeat and revolution sharpened and accentuated the redemptive message of the airplane. Moreover, the very lack of a unified consensus amongst tsarist officers regarding doctrine, mission, or self-identity ensured that potential resistance to aviation would be neither monolithic nor overwhelming. This fluid context offered invaluable space and opportunity for the early proponents of aviation to articulate their visions and pursue their ambitions. It also to an extent ensured that the initial impact of aviation upon the officer community as a whole would be limited.

Yet the inclusive message of initiative and voluntarism would continue to attract officers like Nesterov. Frustration with inertia, red tape, and routine in the face of efforts at reform was a major theme in memoirs of this period; the airplane instead offered a symbol to those in the army wishing a more modern military and prosecution of war, and to those who believed competition with the West must be taken on the same terms and not platitudes. Aviation thus offered a unique forum of activity for officers both juniors and senior: these were the officers who flocked to the aviation schools, who joined the aeroclubs, and who took it upon themselves to organize measures to see that the promise of the airplane was fulfilled. Together these officers comprised yet one more constituency devoted to the airplane, a constituency with its own varying motivations, priorities, and goals.

Equally important were the common visions and hopes shared by the diverse enthusiasts of aviation. Quickly recognizing the power and potential of the airplane, its proponents confidently offered predictions of the airplane's present and future capabilities and harkened to what could be achieved with the right people at the controls. At the same time they also expressed anxiety regarding the consequences of delays, missed opportunities, and the disturbing possibilities of the airplane falling into the wrong hands. Nevertheless, their belief in the boundless future of the airplane and the rapidity of its development combined to create a faith that the bright horizon of tomorrow was ever closer to the present.

These visions, hopes, and fears were usually articulated around several overarching themes. These included the faith in aviation as a technology whose greatest benefits were to come in the future, but whose immediate benefits were to be military ones; the question over whether the most rapid and effective development of aviation would come through the utilization of foreign or domestic technology, expertise, and industry; concerns about what the demands of aviation would entail for established structures, constituencies, and customs within the officer community; and the simultaneous recognition of and anxiety about the necessity of civilian aid and assistance in the development of military aviation. Such themes revealed cultural biases, social concerns, and political attitudes unique to the tsarist officer corps as a community, while also revealing that community to be integrally connected to the broader social and cultural milieu of the last years of the Empire. Moreover, these themes would continue to dominate the discourse surrounding aviation development in tsarist Russia throughout its brief history, prefiguring later responses and policies. The army's early encounters with the airplane were thus instrumental in shaping the structures, attitudes, and expectations that were to guide the subsequent development of the Imperial air forces. Similarly, they largely established the parameters of debate and criticism amongst later generations of historians. Lastly, these initial responses, even among aviation's champions, further betrayed an ambivalence regarding what the onset of aviation portended for the future of the military and the Empire itself. This ambivalence was to shadow the brief history of Imperial Russian aviation to its end.

3| Paths to the Future

In the spring of 1911, Aleksei Poltaratskii, a somewhat regular contributor to *Razvedchik* on aeronautical affairs, commented at length upon the International Aeronautical Exhibition being held in Moscow. Observing the quality and rapid progress of the aircraft on display, he observed that

> to us, Russians, it is especially pleasing that, having quickly taken a distinguished place by the skill of our pilots (I will not even say anything about their courage), we are appearing with distinction also as designers of airplanes, although they do not represent *original* national inventions, in the way of the sad memory of the secret device of Tatarinov or those five airplanes which the officers of the aeronautical park immediately began to build several years ago, having preferred to 'invent' and not learn.[1]

Poltaratskii's mild barb highlighted one of the most significant and contentious debates within the Russian aviation community, the most effective way to build an air arm. Unlike the other great European powers, the Russian Empire simply did not possess the requisite industrial and economic base to manufacture and service aircraft on a large scale, nor would such capacity be available in the immediate future.

As noted in the previous chapter, in light of such prevailing realities and the unique dilemma they presented, prominent Russian aviation enthusiasts proposed two competing solutions. Some argued that, at least in the short term, the Empire must look primarily abroad, to foreign industry, for the purchase of requisite aircraft and related technologies, while others advocated stimulating and supporting the nascent domestic aviation industry with subsidies, preferential contracts, and related financial assistance.[2] The first camp was again prominently represented by the Grand Duke Alexander Mikhailovich and his quasi-public organization the Department of the Air Force, counting senior military officers, state officials, and public dignitaries among its members. The second was led by several prominent aviation officers, various War Ministry officials, and their civilian supporters, with their most vocal champion being General A.I. Kovan'ko, long-time commander of the Officers' Aeronautics School. Their respective arguments drew upon the army's own experiences of the dilemmas of military modernization as well as a far broader history of Russian attempts to define "progress" and overcome backwardness. As such, from its beginnings the Russian encounter with the airplane was self-consciously framed in terms of a unique national endeavor, juxtaposed to the experience of the nations of Western Europe. Within this discourse the army, the airplane, and Russian national identity were inextricably intertwined.

From his lofty perch, the Grand Duke argued forcefully and repeatedly that the new Russian air force must be built primarily with foreign technology. In pointed contrast to the ill-fated experiments to construct aircraft at the Aeronautical Training Park, he decided immediately to defer to the expertise and experience of foreigners, choosing both to purchase aircraft and initially train officers abroad, primarily in France, the leading aviation power. His guiding conviction was that national pride should not blind one to national need, notably the need to build an air fleet both quickly and successfully. To him it was quite clear that the Russian Empire simply did not possess the manufacturing base, the technological infrastructure, or the reservoir of technical knowledge and experience to construct airplanes on a large scale. Nor would the Empire possess such a capability any time soon, at least in his opinion. Thus, it was foolish to throw great sums of money at the fledgling domestic aviation industry in an attempt to spur its development, as the handful of Russian industrialists involved in the matter advocated.[3] This stance did not preclude, of course, buying quality equipment domestically. Based upon his naval experience, however, the Grand Duke decided that the best immediate use of limited public monies would be foreign orders.[4]

In the eyes of the Grand Duke and his supporters, the dictates of military might and national power had to be the only considerations. Satisfying those dictates, at least in the field of aviation, clearly meant that one must look abroad. In like fashion, appropriating technology and expertise abroad did not indicate any subservience to the West, but merely a clear recognition of prevailing realities and necessities. In his public appeal included at the beginning of the 1913 booklet *The Air Force: The Strength of Russia*, published by the Moscow Committee for the Collection of Donations for the Air Force, the Grand Duke noted that in all 77 aircraft had been acquired abroad or domestically in two years by the Department of the Air Force.[5] He expressed his primary concern in his conclusion:

> it is not subject to doubt, that if we do not make every effort for the development of that which is already begun, we will lose that place (second in the world) which we now occupy in regard to an air force, and we will fall behind our neighbors. The danger of the situation is clear, it cannot be allowed, and I again consider it my duty to appeal with a call for donations to the air force.[6]

It was an appeal that was repeated in newspapers and journals throughout the Empire; a good deal of the donated funds would find their way to France, the acknowledged leading aviation power and not coincidentally Russia's most valuable military ally. Again, however, the Grand Duke's methods did not entail obeisance to the West, merely judicious appropriation of its expertise; they signified not a treasonous denial of native Russian genius and abilities, but the most enlightened form of patriotism.

Invaluable support for these policies was provided by the Empire's own aviation hero, Mikhail Efimov. Regarding the need to acknowledge and utilize the expertise of the advanced Western powers in order to ultimately best them, he was thus of the same mind as the Grand Duke. Typical was a report of Oct. 7, 1911, to the War Minister, in which General Aleksandrov, commander of the Main Engineering Administration, explained that airplanes and related equipment constructed in Russian factories were still generally recognized as unacceptable for military use, "and the aviator Efimov especially and insistently pointed out the necessity of obtaining motors and the propellers for them from abroad."[7] Lest anyone still doubt, further praise of the Grand Duke's methods was provided by the renowned former civilian aviator N. E. Popov, who had helped train the original cadre of aviator-instructors of the Officers' Aeronautics School.[8] In his propagandistic brochure of 1912, he evocatively entitled a chapter "The Word of Peter the Great," quoting the Great Reformer to assert that any delay in providing for the defense of the country with this new and powerful aerial weapon would be akin to "untimely death."[9] He especially praised the foresight, energy, and efforts of the Grand Duke, "who had followed in the footsteps of his great ancestor" and had recognized Russia's need for a powerful aerial fleet, as Peter had recognized the need for a naval one.[10] Exhorting the public to greater donations and proposing to establish prizes for successful flights, he admitted that these prizes would be granted "to Russian pilots on planes built in Russia from Russian materials, merely with French engines, owing to the lack of our own as yet." Reiterating the flexibility of the Grand Duke's message, however, he emphasized that engines too should be built in Russia.[11]

An alternative model of aviation development and civil-military cooperation was provided by Colonel, later General, Kovan'ko, engineer, aeronaut, and longtime commanding officer of the Officers' Aeronautics School. Above all, General Kovan'ko was an eloquent and persistent advocate of the view that the success of any aviation program necessitated the mutual and harmonious efforts of military, state, and civilian figures and institutions. Given the airplane's novelty, rapid progress, technological sophistication, and industrial demands, only such multi-faceted cooperation would allow Imperial aviation, civil and military, to reach its fullest potential. In an interview for a newspaper article entitled "The Airplane in Future War", Kovan'ko expressed his certainty that Russia, having already produced excellent pilots in a short time, would turn out as well a series of talented inventors of aviation technology.[12] He thus stressed the fundamental importance of developing a domestic Russian aircraft industry, in terms of materials, design, and labor.[13] As if to underscore the urgency of his appeals, the article in question also included interviews with Mikhail Efimov and A. I. Guchkov, chairman of the Duma's Defense Committee and a figure avidly interested in maintaining the military power of the Empire.

General Kovan'ko provided an even clearer expression of his philosophy and vision in his pamphlet *Thoughts Regarding the Planned Organization of Military Aeronautics in Russia.*[14] Noting the "enormous significance" aeronautics had begun to acquire in affairs of state, industry, and especially the military, he emphasized the pressing need to develop a rational, comprehensive program to utilize effectively all civilian and military personnel, resources, and institutions in this field. As such, the pamphlet contained a detailed organizational plan for the broad development of military aviation and the coordination of the necessary activities of the relevant civilian and military organs. Kovan'ko personally sent a copy, along with an explanatory cover letter, to the War Minister himself,[15] but by the very fact of its publication it was also a propagandistic tract aimed at influencing a wider audience. Kovan'ko expanded upon these themes in an essay entitled "The Air Force" which he contributed to the afore-mentioned booklet *The Air Force: The Strength of Russia.*[16] This was yet another publication which demonstrated the very public nature of the development of military aeronautics in Russia as well as in Europe. In his essay Kovan'ko pointedly addressed the very status and place of Russia in the world political, economic, and social order. He observed that the Empire's enormous expanse necessitated control of its air space as well, and "for this is necessary a powerful air force." This was easily achieved by smaller European states, which possessed advanced technical resources as well as dense, well-equipped networks of railroads and highways. These were blessings, Kovan'ko lamented, which Russia still lacked: "our distances are huge, our roads few, and the ones we do have poor. Looking ahead to the already near future, we can say that Russia needs an air force not only as a means for battle with enemies, but especially as a means of communication."[17]

While European states were blessed with certain natural advantages, Kovan'ko nevertheless identified the real source of the successes attained abroad: "society and commercial circles in Germany set themselves the patriotic task to create a first class, if not first in the world, military air force," while "success in the cause of aeronautics in France is considered by every true Frenchman their sacred duty and the most patriotic affair." In both nations, evidence of this civil-military partnership was again provided by the large public donations for the air forces, numerous factories dealing with aeronautics and aviation production, and the widespread training of both military and civilian pilots in private aviation schools.

In Kovan'ko's eyes, if the industrial and economic conditions of Western Europe were hardly applicable to the Russian context, this patriotic partnership could and should be copied and applied to the unique demands of the Empire:

> we need, in a word, to build ships of Russian design, adapted to the distinctive features of Russian territory. Foreign ships are hardly suitable for us. Airplanes we are already building, engines we are starting to build at home, flying ships we can also build at home. It is a matter only of good desire, of trust in the Russian people and the Russian genius and an absence of the fear of suffering failures.[18]

Kovan'ko then proceeded to the heart of the matter: "what indeed does Russia need in order to occupy one of the first places in aeronautics among other great powers?" His answers were numerous, but are worth quoting at length:

> I. It needs that the creation of a Russian military air force becomes the patriotic duty of every Russian.
> II. It needs that that we create here all that is necessary for the manufacture of aeronautic al devices at home, by Russian minds, by Russian hands, and from Russian materials.
> III. It needs that we not fear failures, but persistently proceed towards the goal of creating our Russian school of aeronautics, our devices and systems, and not go in the tow of foreigners.
> IV. It needs that not only the war ministry and idle bureaucrats be interested in aeronautical affairs, but that the entire people develop an enthusiasm for aviation.
> V. It needs that private persons donate money to the creation of an air force, and that private aeronautical organizations develop widely in Russia, and in practice demonstrate everywhere the successes of the Russian air force, as Zeppelin did, flying over all of Germany, where, it seems, there is not one inhabitant or even one infant who did not see Zeppelin's airship or who doesn't know what a zeppelin is.[19]

Kovan'ko flatly declared that "Russia needs an air force more than anyone." He punctuated his exhortations with his own invocation of the name of the Great Reformer:

> by the precepts of Peter the Great, it was necessary to learn to work as a carpenter and to operate sailing ships abroad, but the fleet was to be built at home. We already know

> how to work as carpenters and how to operate aerial devices, and we should make everything at home. At the present time contemporary air forces, in comparison with future ones, cost as much as did the ships and fleets of Peter compared to our contemporary naval fleet of dreadnoughts.[20]

With this in mind, Kovan'ko concluded that "we cannot lose valuable time, we must stubbornly and insistently propagandize the patriotic matter of the creation of a powerful Russian military and private air force. In it is the power, strength, and future of Russia."

Here was General Kovan'ko's vision of aerial might, presented in detail for all patriotic Russians who were interested in the cause. Most striking was his refusal to regard the development of military aviation as a strictly military problem, which indeed provided the motivation and rationale for publishing his views. Based upon his personal beliefs and professional convictions, he instead emphasized the need to see the development of aviation as a grand national project, one that would require visionary planning, unshakeable determination, unceasing effort, and the active cultivation of domestic reservoirs of genius and productive capacity.

General Kovan'ko's philosophy of aviation development thus centered upon the principle of national self-reliance, construed in its broadest sense. This principle, stated in no uncertain terms, served to clearly distinguish his proposals from the prevailing policies of the Department of the Air Force. The Grand Duke envisioned a more limited role for the public, essentially demanding from civilians only donations, and in particular was far more receptive toward the utilization of foreign resources. In contrast, for Kovan'ko, Russia necessarily had to follow its own path to a mighty aerial future, and the journey had to begin now. Setbacks would indeed occur, but surrendering to domestic backwardness and looking abroad for technology and expertise, even if only temporarily, would merely perpetuate Russian weakness. Emblematic was Kovan'ko's own invocation of Peter the Great, the unquestioned archetypal figure for all participants in the great debate regarding the meaning, form, and cost of modernization in Imperial Russia.

The use of Peter as icon and validation was purposeful and pointed, given the Reformer's titanic stature and resonance in the great debate over the course of Russian modernization. Yet despite the pleas of both the Grand Duke and Gen. Kovan'ko for the active and enthusiastic support of the civilian public, their competing models of aviation development both also implicitly spoke to a crucial feature of late Imperial military culture: the frequently tense, even hostile, relationship between the tsarist officer corps and Russian society in this period. A major source of this tension was the perceived clash of values between those of the military environment and the civilian sphere, between a culture of duty and service and one of self-interest and egoism. Among officers, this tension was manifested strikingly in an abiding hostility to modern industrial capitalism and its attendant values. Such hostility had broader and deeper strains within Russian society and culture; it drew upon a well-established vocabulary that denigrated the materialism, selfishness, and rampant individualism of the West, in contrast to the selflessness, high-mindedness, and collectivism that allegedly prevailed in Russia.

Such anti-capitalist inclinations were uniquely provoked and exacerbated by the nascent technology of the airplane. Throughout Europe, entrepreneurial ambition and commercial considerations governed much of the aviation activity of inventors, pilots, financiers, and businessmen, with the frequent support of military officials. However, while undeniably successful in spurring aviation development abroad, such a state of affairs was a source of notable suspicion among tsarist officers. As such, throughout the short history of Russian Imperial military aviation, anti-capitalist sentiment exercised a powerful influence. Broad appraisals of the airplane's significance, comprehensive projects of aviation development, and specific policy decisions regarding technology and personnel were all framed through the negative prism of anti-capitalist, and by extension anti-Western, suspicion.

As discussed in chapter one, given the numerous fractures and divisions within the officer community, its few unifying social and cultural aspects were often ones, again defined by Fuller as "negative corporatism." Narrow though it might be, the worldview of most tsarist officers was defined by social and professional exclusiveness, a corporatist mentality framed around selflessness and service, and a latent hostility toward the non-military sphere; their values included an exaggerated sense of honor and duty, a conservatism that relished tradition and protocol, and a denigration of civilian pursuits. It would thus be difficult to imagine a worldview more hostile to the fundamental tenets of capitalism: competitive markets, entrepreneurial ambition, the incentives of profit, the acceptance of economic and technological change as both inevitable and progressive, individual attributes of rational self-interest, and the calculated exploitation of opportunity and, if necessary, fellow man. Such suspicion of capitalist vales was certainly not unique to the Russian military; it was a trope amongst contemporary German, French, and British officers to contrast their noble calling with the selfish, money-grubbing, "green beize routine" of the realm of business, industry, and finance.[21] These were cultures, however, where capitalism was established as a successful economic model, where capitalists were thriving, and where a developed bourgeoisie had achieved some level of economic independence, class-consciousness, social and cultural acceptance (if grudging), and political influence.[22]

Such was not the case, however, in the Russian Empire. Russian officers drew their anti-capitalist attitudes, if exaggerated, not merely from the exclusivity of the military milieu, but from a much broader discourse. As various contemporaries and historians have attested, anti-capitalist sentiments were deep-seated in Russian culture, shared by virtually all social classes and groups, and stubbornly resistant to change. James West succinctly notes that negative stereotypes of Russian merchants and capitalists "resonated with and reinforced a deeper prejudice embedded in the public mind"; "individual initiative, private entrepreneurship, and profit-making thus suffered near universal opprobrium, with predictably negative consequences for the economic development of the country."[23] Timothy McDaniel notes that this was in part because fundamental concepts and institutions essential to the legitimacy of capitalism—respect for private property, rule of law, contract theory—were poorly developed in Russia due to its autocratic political structure. Even state officials who accepted the indispensability of private entrepreneurial activity inherently distrusted the spontaneity, unpredictability, and self-interest—all beyond control and tutelage of the state—that characterized contract and market relations.[24]

In his comprehensive study of the history of Russian capitalism through the tsarist and Soviet periods, Thomas Owen emphasizes the continuity of anti-capitalist sentiment as a significant component of a more generalized Russian xenophobia: "then and now, corporate capitalism was hated in Russia because it was foreign."[25] Owen further identifies

> three closely related phenomena in Russian culture: hostility toward capitalism based on a low regard for behavior essential to success in the marketplace, such as calculation, saving, deferment of gratification, and the like; the impulse toward economic nationalism, which at times lacked a specifically xenophobic component; and the xenophobic rejection of capitalism… all three attitudes reflected the sense of a deep cultural gulf between Russia and the West.[26]

What Owen calls "Slavophile capitalism" was in part a source of such attitudes and in part a response. A theory articulated by native Russian merchants to justify and defend their anomalous position in Russian society, "Slavophile capitalism" had distinct elements: it was allegedly more noble, benevolent, and paternalistic than its Western European variant; it drew upon a culturally deep-seated economic xenophobia; and it emphasized the notion that domestic Russian industry was crucial in defending the Empire from the economic imperialism of the West. Russian capitalists could thus assert that their "defense of Russian industry proceeded from patriotic motives, not love of the West."[27] As Owen observes, however, Slavophile capitalism proved to be a tough sell outside its heartland in Moscow. Ironically, in part because of such failures, native capitalists were not only resented for what they did, but for what they failed to do. As McDaniel notes, "public sentiment in Russia added several additional indictments connected with the relative backwardness and dependency of Russian industry: that the industrialists were not dynamic enough, that they were an artificial creation of the state, and that they were in league with foreigners and betraying the interests of Russia for their own gain."[28]

Finally, anti-capitalist sentiment was not merely a set of economic prejudices. As Owen emphasizes, rejection of foreign capitalism had a powerful moral, even spiritual, component; it reinforced tendencies to juxtapose and fetishize Russian cultural superiority, despite economic backwardness, with the economic and technological superiority, but spiritual poverty, of the West.[29] It was thus an integral part of a broader cultural heritage, shared by various groups across the social and political spectrum- workers, peasants, radical and reactionary intellectuals alike- that continued to distinguish Russia from other civilizations, the West in particular. Particularly in the late Imperial period, in a time of increasingly rapid and disruptive social, economic, and political change, anti-capitalist sentiment remained a crucial and unifying component of Russian national identity.

Historians such as Scott W. Palmer have noted the importance of such anti-capitalist attitudes in influencing the Russian development of the airplane; in his comprehensive study of aviation culture through the tsarist and Soviet periods, he identifies anti-capitalist sentiment as one distinguishing characteristic of a uniquely Russian variant of "air-mindedness."[30] Yet such generalized anti-capitalist attitudes and their xenophobic tendencies were particularly augmented and exaggerated within the specific context of *military* aviation, where martial corporatism, deep-seated cultural values, and nationalist insecurities were fused with the imperatives of military competitiveness and state defense.

Such sentiment was also a response to the unique conditions of aviation development abroad. In one sense, the early history of the airplane was characterized by a carnival atmosphere throughout Europe. Stunt and exhibition pilots regularly performed on a circuit of public exhibitions, festivals, and contests, attended by fawning crowds, prominent dignitaries, and interested government officials. Such spectacles also drew the curiosity of adventurers, poets, artists, and intellectuals, eager to participate in or at least document the novel and thrilling phenomenon of human flight.[31] In a less lofty sense, it was clearly a commercial and entrepreneurial orientation that underpinned much of European aviation development. Emblematically, soon after their successful test flights, the Wright brothers quickly sought to convert their triumph into financial gain; they were notoriously hard bargainers, jealous of their patents and conscious that their largest potential customers were European militaries.[32] As airplane fever caught on amongst the European public, contests, prize money, exhibition fees, and high tuition charged at leading flight schools became characteristic features of the aviation scene, with prominent inventors and industrialists, leading pilots, and wealthy patrons the prominent players.

Moreover, As John Morrow emphasizes, Western European militaries were intimately involved in shaping this system through what he terms the "militarization" of aviation, in which military and civilian authorities collaborated patriotically to ensure that the priorities of military aviation, not civil, would prevail. Specifications for both equipment and pilot training were standardized to suit military preferences, while the manipulation of contest rules, prizes, and subsidies to civilian aviation organizations served to dissuade development of sport and other types of aircraft. Through the promise of material gain, this system was supposed to, and to a great extent did, facilitate advancements in aircraft technology and piloting techniques that would ultimately reap military benefits.[33] As noted, however, for many Russian officers such an explicitly capitalist model of aviation development was regarded with suspicion and contempt. Reinforcing the synergy between anti-capitalism and xenophobia, in this view the Germans and even allies like the French could play the part of foreign "exploiters" of the noble cause of aviation. Yet the immediate challenge was to find an effective response to this model. For Russian officers, unlike courtiers, intellectuals, and even many bureaucrats, backwardness was not an option.

Thus it bears emphasizing that both the Grand Duke and General Kovan'ko fundamentally drew upon the same reservoir of Russian anti-capitalism to justify their projects; they merely invoked and emphasized certain strains of such sentiment to suit their respective agendas. Regarding the Grand Duke, in his public appeals for donations for the air force he was addressing implicitly the assumptions and prejudices of his domestic audience.[34] While having few illusions about the motivations of foreign capitalists, he expected that the public would share his assessment that native Russian industrialists necessarily lacked the skill, experience, and wherewithal to compete immediately in such an advanced technological endeavor as aviation. Such an assessment was to an extent unfair; David Jones notes that despite tremendous obstacles and essentially starting from scratch, the domestic Russian aviation industry was showing remarkable progress by 1914.[35] However, the history of the Russian business sector's backwardness, inferiority, state dependence, and essential "foreignness" ensured that the

Grand Duke's position would be both plausible and widely shared. Indeed, the specific failings of the Russian domestic aviation industry—lack of initiative, limited production capacity, poor quality of manufacture—were the targets of frequent and often vociferous criticism from the press and the public.[36] Further, not surprisingly, the few Russian factories involved in the manufacture of aircraft were themselves founded mostly by French firms with French money, relied heavily upon the skills and services of foreign technicians, and built mostly licensed copies of French designs.[37]

Advocates for reliance upon domestic aviation production, most notably Gen. Kovan'ko, also implicitly acknowledged domestic anti-capitalist sentiment, despite seeking to harness it in a very different policy direction. In his numerous public interviews and writings, while referencing the aviation achievements of foreign nations such as France and Germany, he purposely and systematically sought to downplay financial considerations, identifying patriotism, *not* profit, as the driving motivation uniting military and the civilian public. By doing so he promoted these European models of civil-military partnership as suitable for Russia.[38] Further, in putting a proud face on domestic Russian achievements, Kovan'ko also addressed the prejudices of a domestic audience, particularly the deep-seated distrust of and aversion to capitalists among broad strata of the Russian populace. Implicitly, but consistently and forcefully, Kovan'ko made it clear that the motivations of domestic industrialists could be trusted, that their skills and efforts were resources to be utilized, not despised. In his vision of a uniquely Russian model of aviation development, Gen. Kovan'ko thus appealed to and appropriated the most patriotic assumptions and optimistic promises of Slavophile capitalism, magnified before the evocative symbol of the airplane.

In like fashion, in the eyes of skeptics, both camps could be hoist on the petard of their own anti-capitalism. In light of the history of Russian industrialists, who bore the dubious distinction of being both avaricious *and* incompetent in the mind of the public, it was easy to view Gen. Kovan'ko's appeals to domestic production, even if well-intended, as premature, impractical, and risky, particularly with an eye to the imperatives of military competition. Even those Russian inventors and industrialists who came forward with promising designs and innovations often admitted that they lacked the means and capacity to convert those breakthroughs into mass production, at least without large monetary advances and subsidies.[39] On the other hand, as Jones emphasizes, "dissatisfaction with foreign machines and servicing remained a constant theme" throughout the short history of the tsarist air service.[40] Officers like Kovan'ko warned of the danger, and implicit humiliation, of over-reliance upon foreigners, particularly in light of the tendency of unscrupulous foreign capitalists—even those ostensibly "allies" of the Empire—to hoodwink and bilk the Russian state and its people in the pursuit of their own profits. This was again a suspicion and criticism frequently voiced in the press, both civilian and military.[41]

In response to this dilemma, the tsarist military leadership sought to split the difference. Scott Palmer is largely correct that in effect the army, and particularly the Department of the Air Force, ended up following a "buy now, build later" policy. However, in preference and practice, the War Ministry pursued a dual-track strategy of buying technology abroad while trying, whenever possible and practical, to promote industry and encourage innovation at home.[42] Yet despite a general policy of favoring and abetting domestic production, anti-capitalist sentiment ensured that a certain ambivalence regarding such efforts persisted. On the one hand, as a rule the War Ministry did what it could in response to Russian inventors, industrialists, and public organizations who regularly petitioned for contracts, subsidies, and extraordinary funds for events such as festivals and contests.[43] In an October 1912 report to the state Duma, War Minister Sukhomlinov noted the efforts of domestic aircraft builders, yet invoked the Slavophile capitalist formula of patriotism, economic nationalism, and state assistance by warning that attracting domestic capital to the development of aviation would only be possible if there was "the hope of government support of the cause, able to defray if not all, than at least part of the expenses" incurred in aircraft production.[44] Likewise, in a report of October 1913 Chief of the General Staff Zhilinskii argued that purchasing Russian designed and built Kalep engines "moreover would serve in part to encourage domestic industry."[45]

Yet, military officials would continue to have justified apprehensions regarding the abilities and promises of domestic industrialists and inventors.[46] An October 1913 report of the Aeronautics Unit of the General Staff fretted that the three primary domestic aviation firms would be unable to fulfill current War Ministry orders in a timely fashion, to say nothing of the large aircraft orders envisaged in the near future with the expansion of the air services.[47] Further, if the Grand Duke questioned implicitly the experience and competence of domestic aviation entrepreneurs, military officials were at times skeptical regarding the underlying motives and sincerity of civilian petitioners. A press report from the Third All-Russian Aeronautical Congress in 1914 noted that in his presentation the prominent entrepreneur S. S. Shchetinin of the First Russian Society of Aeronautics had provoked "great debate" by criticizing the War Ministry policy of dispersing orders among various domestic firms to encourage competition, arguing for stricter War Ministry controls upon Russian firms assembling imported aircraft components under the guise of "domestic production," and arguing for War Ministry assistance in the form of not prizes but monetary advances, in order to free industrialists from overpaying "foreigners and financiers." [48] Disinterested, patriotic advice, or a desire to corner a potentially lucrative domestic market?

Moreover, the successful development of aircraft technology did not depend solely upon the physical construction of an air force; it also demanded the exploitation of human resources, in the form of engineers, designers, inventors, mechanics, and particularly pilots. Here again considerations of domestic versus foreign resources, and the broader valuations embedded within anti-capitalist sentiment, would continue to frame choices and responses, in both positive and negative ways, for both the Grand Duke and his OVF and Gen. Kovan'ko and the army leadership. In this sense, however, the differences between the two camps were somewhat muted, as shared assumptions produced fundamental agreement about the desirability and limitations of civilian assistance in the cause of military aviation.

Once again, the harmonious union of military and public was emphasized and preached above all. For example, the small cadre of Russian civilian aviators was mostly well known to the officers of the aviation service, and they along with domestic inventors often worked closely with the military testing new equipment and ideas. The noted aviator N. E. Popov had in fact helped train the original staff of the Aviation Section of the Officers Aeronautics School.[49] Likewise, in moments of both celebration and tragedy, the union of Russian military aviators with their civilian counterparts and Russian society at large was celebrated in public rituals and remembrances. At the First All-Russian Festival of Aeronautics in fall 1910, the handful of Russian military and civilian pilots participated together in a popular and patriotic display of the power and potential of this awesome new technology. Russian military and civilian pilots were thus perceived to be imbued with and bound by a sense of common purpose and ethos which distinguished the nobility of all-Russian efforts to conquer the air from the avarice, crude self-interest, and crassness which reigned in the skies of the capitalist West.

This contrast was again to a great extent personified in the person of Mikhail Efimov. In his praise of Efimov noted earlier, Gen. Kovan'ko pointedly emphasized that "he does not resemble those foreign racing drivers, who fly only for the sake of money;" in contrast, Efimov and his fellow Russians were motivated by bravery, selfless devotion to their cause, "and the complete absence of monetary considerations."[50] Efimov duly echoed such attitudes in his analysis of the numerous aviation calamities throughout Europe in the years before World War I. He noted that "explaining this is very simple: the factory owners who build aircraft are cunning, thrifty, and crave making money. For this purpose they invite as mechanics poor, clever, active, and fearless young people, teach them how to fly, and enter them into contest-competitions, where, flying, they bring great benefit to the manufacturer and experience for themselves." He added that if such pilots did not perish from the difficult conditions of many of these contests, their employers then often gave them untested and experimental aircraft to fly. Indicting the whole system of employers and hirelings, Efimov noted that such pilots longed to fly on the best airplanes, but "material insecurity" forced them to fly and die on such dangerous craft.[51]

However, although Russian civilian pilots escaped much of the severity of the criticism directed at foreigners, and indeed earned genuine respect for their efforts to advance imperial aviation, their function was still to perform, to entertain, and to compete for money and prizes. This was not the work of officers. For example, the participation of military pilots in civilian organized flying exhibitions and contests was a common practice in both France and Germany. Such events were designed to encourage the general development of aircraft technology and piloting techniques, as well as specifically military applications. In Russia, the War Ministry had allowed officers to take part in the First All-Russian Aviation Week in My 1910 as well as the First All-Russian Festival of Aeronautics later that fall, although it forbid them from accepting any monetary prizes. Subsequently it ruled upon similar participation on a case by case basis, usually in response to petitions from civilian organizers. While such organizers usually cited their patriotism and respect for the military in their requests, officer-pilots were also regarded as an excellent draw for paying spectators.

Ambivalence and apprehension regarding such participation thus persisted within military circles. A revealing case arose in early 1913. The courtier Prince Abamelek-Lazarev had established an anniversary trophy in honor of the 300th anniversary of the Romanov dynasty and offered it as a prize for the winner of a flight between Moscow and St. Petersburg. The flight was to be organized by the Imperial All-Russian Aeroclub, which had sponsored the first such flight in 1911, in which thirteen civilians but notably no military pilots had participated.[52] In a letter to Chief of the General Staff Zhilinskii, Prince Abamelek explained that his goal was to encourage Russian military pilots to carry out the types of tasks they would encounter in time of war; he thus appealed to Zhilinskii to allow officers to both participate and be eligible to receive the prize, as officers did in France, Germany, and Austria during the aviation weeks there.[53] In Zhilinskii's subsequent report to War Minister Sukhomlinov, he noted that "in light of the interest to military aviation" that the proposal might present, he was submitting it for approval; he added, however, that if such participation was allowed, it should be done "at the officer's own risk, without any interference and expenditures of the War Department."[54] The War Minister did indeed decide to allow two military pilots to participate; the Aeronautics Unit of the General Staff subsequently issued a circular letter requesting lists of officers wishing to be considered.[55]

A rather negative reaction was provided, however, by none other than Gen. Kovan'ko. In a report to Gen. Zhilinskii, he noted that among the officers of the Officers' Aeronautics School, "where pilot-instructors are continuously busy," there were none wishing to take part in the proposed flight.[56] He observed that in contrast to pilots of the aviation squadrons, "the conscientious performance of the serious duty of an instructor" would make participation impossible. Kovan'ko went on to cite a less obvious consideration: "a military pilot, especially an instructor of the school, who has expressed the desire to carry out the flight will have to obtain the prize at all costs, since if from all of the officers who wish to fly the two best will be selected, then they must by their flights maintain the renown of the entire corps of military pilots." At least according to Kovan'ko, however, this was nearly impossible given the conditions of the contest, which were far too disadvantageous for officers. The aircraft flown by the military pilots would be far slower than those of their civilian competitors, who would also be supported with assistance from their respective aircraft firms in the form of spare parts, motors, and mechanics.

Kovan'ko further noted that if there were officers who desired to enter the contest they had probably not weighed the circumstances he had enumerated. In contrast, "the officers of the Aviation Section, and also those who have learned to fly at the aviation section in Warsaw [and] who have not entered the contest, all express their complete readiness to carry out this flight not for a prize, but only for the performance of an official mission." In the opinion section of his report Kovan'ko raised the ante by concluding that "I believe that in general the participation of officers in any competitions whatsoever for money and before the public is not desirable and is in no way justified, since military pilots have the opportunity to receive sufficient and more varied and expedient practice in the performance of all kinds of official tasks." He thus formally requested Gen. Zhilinskii's resolution "to forbid our military pilots from taking part in any competitions whatsoever not organized by order of the command." While it is clear that General Kovan'ko spoke as much or more for himself when reporting the opinions of his officers, it remains a fact that no volunteers from his command came forward.[57]

Such attitudes are particularly noteworthy in light of Gen. Kovan'ko's ardent and insistent advocacy of developing a domestic aviation industry and air arm in close cooperation with public organizations and civilian institutions. They are likewise all the more notable in light of the membership of the Imperial All-Russian Aeroclub. Like most such organizations in Imperial Russia, it was overwhelmingly made up of state officials, prominent civilians, and numerous army officers, including General Kaul'bars, enthusiastic aviation patron and confidant of the Grand Duke Alexander Mikhailovich. Yet the efforts of the Imperial Aeroclub, as well as those a prince in the Imperial court who expressly desired to aid military aviation, were greeted with suspicion and hesitance by the very man who continually appealed to the Russian people to join the army in the great patriotic cause of aviation. For Kovan'ko generalized and abstract appeals to civil-military partnership and national service were stirring and noble; however, the actual exchange of money, the granting of material rewards, and the exposure to the potential criticism of a public seeking a return on its investment reeked far too much of the greed, selfishness, and materialism that distinguished the capitalist inclinations of the civilian, and particularly European, world.

It was not only Gen. Kovan'ko who harbored such doubts regarding the relationship of the military and the broader Russian public, and in particular the critical issue of the work of civilian aviation organizations and enthusiasts. While such civilians always cited their ardent commitment to furthering the advance of Russian aviation, the frequent view of aviation and War Ministry officials was that profiteering, self-interest, and greed truly lurked behind many such requests of these professed "partners." The use of civilian aviation schools for military purposes in particular served as a point of concern. The prevailing model being elaborated abroad, particularly in France, was for civilian aeroclubs and schools to provide basic preliminary flight training for officers; this was then followed by a more intense and specially designed course for military pilots at the military aviation schools. In contrast, from their inception the Russian military aviation schools had largely provided all instruction for officer trainees, primarily because civilian schools were so few, but also due to perceived military imperatives. A report of the General Staff to the War Minister in Nov. 1911 stated directly that "the training and creation of a cadre of reliable military pilots should be exclusively in the hands of the War Department...Therefore it would seem that to resort in this matter to the services of private schools and societies, to say nothing of the possibility of their deleterious influence upon military persons, would seem to be undesirable." Not only would such schools fail to meet the specific demands of training military aviators, but "private aviation schools and societies, pursuing their own particular gains, will give military aviation only as much attention as will be advantageous for them in a given case."[58]

However, despite attempts to expand its own training schools, as the army rapidly expanded the size of the aviation service a shortage of trained pilots increasingly became an issue.[59] As such, the Russian army was increasingly considering the adoption of the French model. Yet desire for military control over the entire training of future pilots, as well as an uneasiness regarding the use of civilian schools, persisted. Such attitudes are again striking in light of the composition of the leading civilian aviation organizations. Moreover, due to the relative paucity of the army's own training institutions and facilities, dismissing civilian assistance in this regard might well be regarded as manifestly counter-productive. However, it appeared that preserving the sanctity and purity of the military sphere from "deleterious influences" might in the end be more important than training a cadre of pilots as rapidly as possible. Ironically and perversely, the suspicions of the War Ministry were frequently confirmed. Palmer notes one notorious incident in which the Imperial All-Russian Aeroclub, after providing substandard training for military officers it had promised to train for free, subsequently reneged on the agreement and demanded an outrageous 60,000 ruble increase to its yearly state subsidy.[60] What better evidence that the self-professed good intentions of civilian entrepreneurs and clubs simply could not be taken at face value?

While the Grand Duke focused his attentions upon donations and technology, some of his colleagues thus proposed their own alternatives. Noteworthy was Colonel Odintsov, an officer of the General Staff, commander of the Sevastopol Aviation School of the OVF, and an energetic and enthusiastic proponent of the development of Russian military aviation. In a short report of June 28, 1912, to the Main Administration of the General Staff, Odintsov laid out what he thought were necessary measures to deal with one of the most fundamental needs of the new and expanding aviation service— cadres.[61] This report is worth quoting in full:

> in all of the armies of the great powers feverish work is going on to increase the aviation forces, which demand a corresponding cadre of pilots, from a young age trained and raised in the spirit of their future service. Proceeding from this, I find inadequate the training of officers sent from the regiments, often already getting on in years, as a result of which I noted an inability to absorb and grasp all of the subtleties of the art of flying.
>
> It is necessary to establish a special military aeronautics school [*uchilishche*], which would accept young people directly from the cadet corps and *gymnasia*. The officers' aviation school would then become a special school for the training of future commanders of the aviation squadrons, and also a station for testing various airplane systems and their application in various cases. The organization of a military aviation *uchilishche* would desirably be carried out in the very nearest future and it would open no later than fall of 1913. The course [of instruction] would be 2 years.

Although the report is rather short, concise, and narrowly oriented, Odintsov nevertheless touched upon some compelling issues. What was clear above all was Odintsov's desire for military control and jurisdiction over the entire training of future pilots, institutionalized in the proposed aviation school. Although the report discussed the need for a special aviation school in terms of the problems of age and ability of line officers to adapt, in this sense Odintsov's proposal can be seen equally as an argument for the retention of all or most training in military hands. At the same time, Odintsov's report implied that in his vision the aviation services would attempt to cull the best people from society at large; the civilian gymnasiums were specifically mentioned along with the cadet corps as a feeder pool for the future aviation services. Although the matter of civilians and reservists was not discussed, in his emphasis upon the "spirit of service" Col. Odintsov implicitly emphasized the distinction between the mores and habits of the military vs. civilian sphere.

However, anti-capitalist prejudices could just as easily be applied in far more pessimistic and doubtful directions. In the January 9, 1911 edition of *Sevastopol Illustrated Aviation Journal*, Colonel Kh. F. Stamat'ev, another future commander of the Sevastopol Aviation School, weighed in with pessimism regarding the ability of civilian pilots to be truly of use to the military in the event they were mobilized.[62] He noted that "without doubt, in wartime there will be heroes," who though not obligated to serve would voluntarily offer their services to the army, but the instances of such offers would be few and their benefit small, "as a military commander will treat very carefully the report of a dilettante." He added that relying on the service of civilian aviators "also does not seem possible, as before the face of obvious death any contract is null and void." Such flyers were in marked contrast to "persons, upon whom life has already made demands and whose role has already been precisely determined, that is a cadre of military aviators." In light of such realities, and as the primary applications of aviation were clearly military, he argued that it was only natural that the majority of pilots should be officers. Stamat'ev could not ignore, however, the concurrent development of civilian aviation. He accepted the need to provide equal training to civilians "interested in aeronautical sport, as from amongst such persons it is possible to advance aviators, who being interested in material advantage in the form of prizes" would introduce new improvements in machines and devices. Stamat'ev went on, however, to warn against the excesses of the commercial impulse. He observed the lure of such material and psychological rewards, noting that the glory of Efimov and other leading Russian civilian aviators and the "colossal salaries" they received "had gone to more than one head" and had "compelled many to quit their everyday occupations, take their savings and one way or the other" get involved in aeronautics. He defined the seductiveness of aviation: "an art which demands neither a multi-year theoretical education, nor much physical labor, and at the same time provides glory and fabulous honorariums- thus or almost thus is how the profession of aviator appears in the eyes of the people at the present time."

Stamat'ev noted, however, that the reality of the matter was rather less romantic and glorious, and that the financial future of performance flying, once aviation was inevitably no longer a novelty, was rather bleak; "therefore the educational institutions, like the Aero-Club, should not close their eyes to this sad occurrence and are morally obligated to open the eyes of the deluded; otherwise the time will not be far off when there will come into being a new type of aerial proletariat, which, unfortunately, is already beginning to be seen." He concluded that such was why the "advertisement character" of filling civilian aviation schools, referring to the common contemporary practice of such schools advertising in newspapers and magazines to attract students, "should be completely cast aside," and the selection of willing students "should be conducted with great care."

Contributors to *Sevastopol Illustrated Aviation Journal* were still warning about this new and unwelcome development several months later. In a lead article entitled "Chauffeurs of Aviation", the author, N. Rozanov, remarked upon the great steps forward of aviation and the heroism of its greatest pilots, such as Efimov.[63] However, he also noted the appearance of a new type of pilot—"airplane chauffeurs." Rozanov summarized the brief career of one V.F. Smit, who had recently died in an airplane crash in Petersburg. He had been a chauffeur at the Russo-Baltic factory; when the factory began to produce aircraft, he had been bound to a five year contract and then trained to fly in Paris. Rozanov noted that in all the time before his fatal flight Smit had spent not more than two hours in the air, and that he had got into aviation in order to earn more money and feed his family.

Rozanov remarked that aviation was evidently entering a new phase, that of a capitalistic enterprise, with hired labor, the exploitation of workers, and the making of fortunes. He observed that "this is natural in the twentieth century. It tears away the mantle of idealism, it replaces heroism with another idea, instead of the joy of victory over the elements" it threatened to bring the laughter of the rich, "bought at the price of the life of the exploited poor." Rozanov concluded that although the deaths of those like Smit made all bare their heads with full respect, "these are no longer fighters for an idea, their death— not the ecstasy of a hero. The chauffeurs of aviation are the product of the capitalist system and their death—sacrifices to Moloch."

What is noteworthy in the warnings of Stamat'ev and Rozanov was the ambivalence regarding the specter that commercialism brought to the otherwise noble and militarily significant pursuit of the progress of aviation and aircraft technology. On the one hand was distaste, bordering on disgust; the lure of monetary reward was nothing as compared to the strength of the role of duty and service, and indeed was largely a pipe dream. In particular, as suggested by McDaniel, Stamat'ev found it impossible to embrace the concept of contract, poorly established in Russia, as truly binding between free and equal partners on the basis of mutual trust. His skepticism was augmented by the clash between military values of duty, selflessness, and sacrifice, and the selfish and base motives which ruled civilian society. However, Stamat'ev in particular recognized that the commercial impulse was a reality, and one that could be and had to be utilized effectively. Despite the palpable lack of trust in those driven by desire for glory, sport, or worse, money, with careful selection those civilians with enough skill, ambition, or greed could benefit the cause of technological advancement, and thus benefit the goals of the military. The alternative to harnessing the desires of these men effectively was clear: a cadre of the flying poor. Moreover, though nowhere stated so explicitly—the language was instead couched in concern for the poor, deluded souls who threw away their time, money, and even lives chasing the chimera of flight—these admonitions hinted at a disturbing vision of a trained, exploited, and even worse, disgruntled proletariat of the air and possibly dangerous technology at its disposal, with all of the potentially nightmarish scenarios that might offer. In this sense anti-capitalist sentiment could transcend economic prejudices and moral judgments, and become instead a vehicle to address issues of social rebellion, political instability, and national security.[64]

Arguments over whether to build at home or purchase abroad, and whether to utilize or resist capitalist tendencies, were not merely academic disputes. Although the Department of the Air Force technically had no jurisdiction over the selection and procurement of official military aircraft types, its financial wherewithal, its Imperial patron, and its semi-autonomous status worked to ensure not only the Department's relative independence in such matters, but even predominance at times. In contrast, while General Kovan'ko too enjoyed no official jurisdiction over aircraft procurement, his command status, long experience, and his officers' work in testing new machines and designs also provided a valuable forum to disseminate his ideas and influence policies. In turn, from Grand Dukes to generals to bureaucrats, from colonels and lieutenants to civilian activists, in published articles, popular pamphlets, official reports, and internal memoranda, suspicion and hostility toward capitalist models and values were remarkably consistent. As such, anti-capitalist bias and its xenophobic tendencies formed an influential cultural context in which officers, often those in positions of authority, assessed policies and made decisions regarding the course of Russian military aviation.

While recognizing the serious constraints presented by the material and economic conditions of the Russian Empire, optimism was nevertheless the rule among officers, framed again through the lens of national identity and social and cultural distinctiveness. With memories of defeat and humiliation still fresh, the redemptive promise of aviation took on heightened appeal for both army and state. The airplane offered greater rewards to Russia than perhaps any other power, offering a means to bypass the accumulated legacies of backwardness. In like fashion, Russia's own impressive aeronautical achievements removed much of the stigma of dependency, and hence resistance, which frequently attended calls to adopt European models of modernization. Lastly, the fruitful partnership of state and public seemingly promised even greater successes in the future.

Thus, at a time when it seemed that the gulf of mutual understanding and respect separating the military and civilian spheres had never been wider, the public's enthusiastic embrace of military aviation appeared to offer a path to that elusive unity of patriotism, the perceived lack of which so often surfaced as a lament among tsarist officers. This civil-military partnership seemingly presaged not a search for alternative paradigms of national identity, but an opportunity to reform and revitalize inherited ones. The powerful visions of renewal, progress, and power invoked in the name of the airplane demonstrated that Russian officers, and the Russian people, could not only rise to the challenge of modernization, but do so in a uniquely Russian, indeed an Imperial Russian, manner.

Such appeals to Russian social and cultural distinctiveness should not be dismissed out of hand as the cynical manipulation of Slavophile cliches, or as the invocation of "magic" to remedy intractable difficulties of technological backwardness and military inferiority. The appearance of such themes in a wide variety of forums—the progressive military press, specialist aviation journals, the broad spectrum of the popular press—demonstrates that they were expected to resonate with a diverse audience of leading bureaucrats and statesmen, senior officers, and the newly assertive civilian public. More importantly, the vocabulary of national identity was used to articulate real and competing programs of aviation development and policy choices regarding the allocation of scarce resources—financial, industrial, and human. One should not overstate the differences between these programs; all of the actors involved pledged cooperation and acknowledged common goals in the great cause of Imperial military aviation. Nevertheless, they identified in subtly different ways the sources and nature of the Empire's social, cultural, and political distinctiveness. They likewise advocated competing approaches to exploit most systematically those national traits which were strengths, and to overcome most efficiently those which were liabilities.

Thus, regardless of his often questioned motivations, the efforts of the Grand Duke Alexander Mikhailovich in the field of aviation represented an effort to reclaim the tsarist mantle of enlightened leadership and stewardship of Imperial power. Quite self-consciously, the Grand Duke appealed to the Petrine legacy of modernization, with the state in its self-proclaimed role as the traditional engine of progress. While the Grand Duke had little doubt that the Emperor's "true" subjects, revolutionaries and restive nationalities aside, possessed the requisite patriotism, loyalty, and yearnings for national greatness, he also believed firmly that, just as under Peter the Great, Russia had to first look abroad for the fruits of industry and technology it did not possess. In turn, General Kovan'ko also appealed to the Petrine legacy in welcoming the foreign cultural baggage the airplane brought with it. In his eyes, it was not Western technology that needed to be appropriated, but precisely the spirit of unity and common purpose possessed by the European powers. Indeed, in his appeals to patriotic self-reliance, General Kovan'ko implied that Imperial Russia's greatest national weakness was not material, but psychological: a lack of belief in the talents, genius, and potential of its own people. He thus advocated not a defeatist retreat to belief in "magic", but a reawakening of the true Russian spirit of progress. It is worth noting, and perhaps a bit ironic for a senior tsarist officer, that Gen. Kovan'ko's diagnosis of Imperial backwardness, and his unyielding advocacy of domestic efforts and national self-reliance, would later win him the praise of Soviet historians.[65]

Yet both sides in this debate shared and were to an extent united by a common anti-capitalist bias. The negative response of many Russian officers to the capitalistic orientation of aviation was not unique. One of the more striking features of the early period of aviation was the yawning gap between the poets and dreamers and the tales they spun, and the largely entrepreneurial and military orientation of aircraft development. Yet in the West, this contradiction was expressed largely in the language of socio-economic class, attendant value systems, and moral opprobrium. Civilian critics of the current state of affairs compared it to a bright future which largely drew upon the language and motifs of socialist and utopian thought, while military skeptics could retreat to the surety of martial values.[66] Moreover, amongst the military and commercial figures who counted most in the scheme of things, patriotism and profit, self-interest and idealism, were not necessarily mutually exclusive.

Russian anti-capitalist attitudes also drew heavily upon moralistic and utopian language, albeit of a distinct variety. Influenced by such broader strains of Russian culture, tsarist military servitors were also conditioned by professionally specific and socially exclusive norms of duty and honor. Anti-capitalist sentiment thus took on a very specific form for such officers. Profit seeking, materialism, and self-interest were contrasted not to socialist ideals, but to patriotism and service to the throne and state as a whole. Thus, for those officers who looked with enthusiasm to civilian assistance in the great project of building Russian aviation, this partnership offered an opportunity to reassert a shared sense of cultural uniqueness, national identity, and patriotic achievement. This awareness was largely predicated upon Owen's typology of a mutual hostility to capitalism, an urge to economic nationalism, and a passive-aggressive xenophobia. For those officers who were more skeptical regarding the motives and intentions of civilians of various stripes, anti-capitalist and xenophobic attitudes also served as a fundamental point of departure. In this regard capitalist values of self-interest and profit-seeking were in and of themselves inherently suspect, irrespective of the particular nationality of the bearer, and of primary importance.

In the face of successful foreign models and the demands of aviation development, anti-capitalist and anti-foreign sentiments could not be given free rein, but they could produce doubt and suspicion. In the context of a broader discourse regarding the relationship between Russian aviation development, military modernization, and national identity in this period, the prism of anti-capitalist sentiment framed the relevant issues and values in particularly stark relief. Ambivalence regarding civilian pilots and public organizations, and particularly native capitalists and industrialists, revealed more than officers' distaste of the contract mentality that seemingly governed relations between military and society in the West. Such ambivalence spoke as much to their assessments of the relative patriotism and devotion to national goals of ordinary Russian subjects. As a result, there was a palpable and continuing tension within the Russian military over the desire for public support and assistance in the development of aviation and the fears of the unforeseen and unwanted consequences that might entail. It would be the Bolshevik regime and its statist commitment to the patriotic cause of Russian aviation that would remove once and for all the ambiguities and unease generated by capitalist models and values.

These debates drew upon and re-energized a more enduring discourse regarding the articulation of a uniquely "Russian" art of war and the nexus of political traditions, social relations, and cultural values that were its perceived wellspring. In turn, attempts to define what was in essence the relationship between military power and national identity were couched within an even older and broader dialogue regarding the place of Russia within the European order. As such, the perceived imperatives and demands of the airplane reflected both the cosmopolitan experience of international aviation while also revealing uniquely Russian conceptions of time, space, progress, and national identity. The immediate challenge was to implement concrete policies and reform established institutions so as to realize the full potential of the Russian air forces.

4| Institutions and Adaptation

The competing programs for the development of aviation proffered by the Grand Duke Alexander and Gen. Kovan'ko involved more than specific choices regarding policies, technologies, and the utilization of foreign or domestic capital and industry. They represented fundamentally different philosophies and visions regarding the future of the Russian air services. As such, they testified to in a negative sense to role of the war department as the single guiding hand in the elaboration of a unified, comprehensive, and coherent policy of military aviation development.

There were those in military circles who believed that such a situation was intolerable. In a lengthy article that appeared in *Razvedchik* in February 1911, the author, N. Dukhanin, provided a detailed critique of the contemporary status of military aeronautics and aviation in Russia. [1] Emphasizing the future significance of aviation as a major factor in warfare, he stressed that despite recent efforts Russia was significantly behind the other great powers in Europe, and indeed was only making its first tentative steps in this direction. Dukhanin stressed that a dispassionate analysis of the situation—itself necessary to avoid mistakes and ensure success—would have to conclude that Russian military aeronautics in essence did not yet exist. Russia possessed neither the necessary supporting technical infrastructure, nor the trained cadres, nor a perfected, comprehensive plan of organization for military aeronautics. He felt it necessary to stress this point, as "in this regard, as in many other cases, there is noticed among us a very large and always harmful tendency to delude ourselves and under this point of view make falsely self-favorable conclusions and deductions, which, however, have only for themselves the one dubious value of the pleasant tickling sensation of pride." He noted instances of such excited self-delusion, notably the numerous press reports of the events of the "All-Russian Aeronautics Holiday" in fall 1910 and the successes achieved at the new Sevastopol Aviation School, which had produced in both military and public circles the misconception that Russian military aeronautics was overtaking European competitors. Giving all due justice to the energetic efforts of "our young pioneers of military aeronautics," he repeated that it was "extremely harmful" to make from this "any pseudo-real and premature conclusions."

Instead, what was needed was practical, intensive, energetic work to ensure the quickest possible development of aeronautics and to allow Russia to catch both her allies and possible future enemies. Necessary first of all was unity of action, a single organization in whose hands would be concentrated all of military aeronautics: doubtless "only such a single organization, a single will is capable of systematically and appropriately coping with this enormous and serious task of our state defense." Dukhanin argued that this should naturally be the War Ministry and its senior leaders, as their task was the preparation and maintenance of Russia's armed forces. However, he pointed to the division of effort represented by the Department of the Air Force, "which, in our opinion, seriously disrupts the unity and wholeness of our military aeronautical organization." Such activity was producing muddle and disarray while "at the same time not having in itself anything organic, stable, or systematic."[2] Dukhanin further claimed that "sympathy for this institution, by the way, is far from shared by all in our military circles." Concerning the successes of the Department, he added that the majority could be explained by the exceptional circumstances in which the organization found itself. These included the prevailing public mood and the elevated interest of the moment, "which, as any mood, is not the result of a thoughtful attitude to the matter and its broad interpretation," the Department's advantageous semi-public, semi-governmental character, its mixed composition, and the significant funds it possessed, themselves constantly augmented by new donations.

Dukhanin observed that the air fleet the Department built was considered the "'property of the donors'" and only in event of war would be transferred to the War Department. He conceded that it would seem such a "completely private enterprise of a public character", through its resources of material and personnel, "could be a good and valuable help to the war department in wartime," but "in reality, however, this is not so." What help it would be in wartime was difficult to say, while that it was now a hindrance to the proper development of our military aeronautics, "in our opinion, can be said with sufficient certainty," as the Department of the Air Force had created its own private aeronautical organization exclusively for officers, namely the Sevastopol School. This school had diverted to itself, in the capacity of instructors, the extremely limited pool of Russian officers trained as pilots: "by this it [the OVF] has paralyzed, in our opinion, nearly all of the aviation activity of the War Ministry and in particular the Officers' Aeronautics School." This alone constituted a significant disruption, "but this is still not all. It inflicts harm on the very matter of the development of *military* aviation." Those officers trained abroad and domestically and now transferred to the ranks of pilot instructors "in our opinion should themselves still be learning to fly, and mainly preparing themselves for the role of *military* aviators." There was "no time to lose" to enter what he considered this second phase of military aviation.

Dukhanin thus concluded that "it seems to us that the founders and organizers of the Department of the Air Force did not entirely correctly look upon their task and broadened it in an undesirable direction." He observed that the only rational sphere of activity for a similar type of organization would be "exclusively active assistance and financial and material aid to existing government and public organizations. But under no circumstances the creation of its own separate organization, which is, in our opinion, an organic mistake which diminishes the fruitful activity of the institution for the good of the state"(67). Thus, the idea of the Department to fund by its own means the training of officers in France was completely correct, but retaining them as instructors was mistaken: "instead of this, the Department should in every way assist and help build and broadly establish under this school [the OVS] a special *aviation section*, which now does not exist, unfortunately, and does not exist, apparently, primarily owing to the parallel activity of the Department of the Air Force." Similarly, he argued that the idea of the Department to supply "our fatherland" with the largest possible number of planes as quickly as possible was proper, but it was impossible to regard as correct the idea of using these planes to create its own aerial flotilla that would be the 'property of the donors.' Instead, he noted that a more worthy role for the OVF would be to use the planes it purchased to supply the Officers Aeronautics School, the aeronautical units, and lastly the leading public organizations, or to help organize a government or a subsidized private factory to produce aviation and aeronautics equipment.

Dukhanin did not stop with his criticism of the Department of the Air Force. He continued with a lengthy appraisal of the various imperfections in the arising organization of Russian military aeronautics, among which "we consider the primary and more fundamental to be the organization of the Officers' Aeronautics School," notably the absence of an aviation section. The soonest possible opening of one, as well as various other measures to improve the selection and training of aeronautical officers, was urgent. Lastly, training was to be extended to all branches of the army. Finally, Dukhanin concluded that "it is necessary to admit that the organization and general establishment of military aeronautics in our army still does not fully satisfy all of the necessary conditions for its rapid and systematic development." While it was true that the matter of aviation had only just come into being, "it is all the more so necessary from the very beginning to eliminate all organizational and other imperfections and also outside hindrances and to place our young military aeronautics from the very beginning of its entrance into a new era of existence on correct and firm foundations" (68-9).

In a short supplementary article three weeks later, Dukhanin revisited the subject.[3] He noted that regarding his previous piece, "it was pointed out to me my insufficient knowledgeability of the role and character of the activity of the Department of the Air Force in the matter of our arising military aviation." In what could readily be construed as less a retraction than a redirection of criticism, he noted that having received factual data, it was clear that for the past first year of its existence the activity of the OVF "did not and could not serve as any hindrance for the War Ministry, as actually during this time there was almost no aviation activity in the War Department at all." Only in the coming year, based on the latest information, would this activity be forthcoming in the form of the opening of the Aviation Section of the Officers' Aeronautics School. He thus applauded the goal of the Department of the Air Force in trying to establish military aviation in Russia as quickly as possible and ticked off the details of the intensive work it had done to achieve this goal.

However, Dukhanin continued that while thus recognizing the positive results of the activity of the OVF,

> I nevertheless remain in the belief that this activity, however productive in and of itself, nevertheless does not fundamentally and radically resolve that important and crucial state task, which contemporary military aeronautics is for our army and its combat preparation. The fulfillment of this task should lie in the direct responsibility of the state itself in the person of the War Minister, responsible for the entire state defense of the country.[4]

He thus could not regard as mistaken his opinion that "the training of the personnel of military aviators should not and cannot enter into the tasks of the Department, as a public organization" and proposed that with the beginning of aviation activity of the War Ministry, this training should be completely in its hands. He reiterated that the tasks of the OVF should exclusively include material aid to the War Department, and concluded that "it is my deep conviction that only under this condition and of course only with the energetic efforts of the War Ministry and the army itself will our military aeronautics embark on the correct path of its development."

Dukhanin's thorough and unsparing critique addressed a number of the themes enumerated above regarding models of aviation development and their application in Russia. Buttressing his entire argument was the recurring theme of the project of "our" military aeronautics and its importance to all of "Russia," presented here as a united, undifferentiated whole, and juxtaposed against the stark backdrop of the leading European powers and their fruitful labors. This patriotic urgency, however, allowed no place for simplistic chauvinism. Instead, Dukhanin pointedly took swipes at what some regarded as a traditional tendency, particularly pernicious in the military sphere, to fetishize Russian culture and things Russian in the ongoing debate of Russia vs. the West, itself largely an argument over the negotiation of foreign economic and technological developments to Russian culture. In Dukhanin's eyes, the gauntlet had already been thrown down, and the standards set, by the powers of Western Europe. In this regard, he also stressed the distinction between the progress of aviation in general, however desirable, and the development of strictly *military* aviation, whose needs and priorities must take precedence.

With these fundamental premises, Dukhanin focused upon those questions that were to prominently engage those officers in leading positions in the aviation services and War Ministry. Above all was the need for a comprehensive, rational, and hierarchical plan of organization for the development of military aeronautics, upon which the successful resolution of both the technical and personnel aspects of the matter depended. In turn, the concentration of all matters of military aeronautics in the hands of the military and its leading institutions was necessary not merely for matters of principle, but to ensure the successful development of military aviation. The alternative was the existing state of affairs: dispersed and uncoordinated effort coupled with inertia and paralysis. At least to Dukhanin, it was abundantly clear that the fault lay with the root of this problem, namely the Department of the Air Force.

We do not know the true extent of Dukhanin's claim that sympathy for the Department of the Air Force was "far from shared by all in our military circles." Yet it is clear that this organization did occupy a somewhat anomalous position vis-a-vis the military establishment. Although proudly and publicly committed to the tenets of military might and Imperial greatness, the OVF was nevertheless engaged in its own distinct project to develop military aviation. While it was the expressed intention and desire of the Grand Duke Alexander Mikhailovich and his subordinates to cooperate in all ways with the War Ministry and the aviation institutions, as we have seen, priorities, methods, and plans were not always identical or harmonious. This was exacerbated by the fact that the efforts of the OVF, as Dukhanin grudgingly admitted in his rejoinder, still dwarfed those of the military establishment, notably the Officers' Aeronautics School. What so vexed Dukhanin was that, in an inversion of the European model, a semi-public organization was to an extent contracting directly with the Russian people to provide for their defense, bypassing those who claimed this as their sworn role and duty: the officers of the army. Although the inertia of the War Ministry was in part to blame, this did not excuse the current state of affairs.

The Aeronautics Unit of the General Staff

Inertia or no, the Russian War Ministry could not ignore the notable advances in military aviation being made abroad, nor its own organizational deficiencies in this area. In the fall of 1912, an important step was taken to deal with the simultaneous lethargy and jumble of uncoordinated activity that characterized Russian military aeronautics and had prompted Dukhanin's complaints. This step was the transfer of the former Aeronautics Section [*Vozdukohplavatel'nyi Otdel*], under the jurisdiction of the Main Engineering Administration, to the jurisdiction of the General Staff. The reconstituted organ was christened the Aeronautics Unit [*Vozdukhoplavatel'naia Chast'*] of the Main Administration of the General Staff [*Glavnogo Upravleniia General'nogo Shtaba*, hereafter GUGSh] and was charged with the general direction of all affairs concerning aviation and aeronautics in the Russian army. This direction involved all aspects of the matter—personnel, hardware and equipment, training, and testing. The Aeronautics Unit consisted of two sections: the first dealt with matters of special training, service, and personnel, the second with technical matters and supply.[5] Moreover, as a testament to the importance being accorded to this new organ and its work, the commander of the Aeronautics Unit, a general of the General Staff, was to report directly to the Chief of the General Staff, who himself was to have overall responsibility for the general supervision of aeronautical affairs.[6]

The specific considerations surrounding the transfer of aviation to the General Staff were laid out by Chief of the General Staff Zhilinskii in his memorandum of June 15, 1911, to War Minister Sukhomlinov.[7] He noted that in a General Staff memorandum of June 13th regarding the inclusion of aviation squadrons in maneuvers, Sukhomlinov himself had placed the resolution "I find it necessary to concentrate the entire matter of aviation in the hands of the Chief of the General Staff" and in accordance with this to transfer all such matters from the Main Engineering Administration. Zhilinskii therefore proposed to abolish the former Aeronautics Section and Aeronautics Committee (concerned with clarifying technical specifications for equipment and evaluating specific systems and inventions) of the Engineering Administration and create corresponding new organs of the General Staff. He noted that as aviation was one of the means of military communications and liaison, these organs were to be placed under the Military Communications Section of GUGSh (83-4). Zhilinskii envisioned that the transfer of the resolution of both organizational and technical questions to the direction of GUGSh would also involve the purchasing and stockpiling of all necessary hardware, equipment, and support infrastructure relating to aviation and aeronautics, analogous to the situation with the General Staff's management of the railroads. Of course this would include control of the requisite credits to deal with such issues, "for only by being complete master in the matter will it [GUGSh] be in a position to carry it out as is required by the state significance of this matter" (84).

Lastly, Zhiliskii proposed the transfer of the Officers' Aeronautics School and its battalion as well to the jurisdiction of the Chief of the General Staff, given its new control over all questions regarding aviation, including the equipping of the army with aeronautical equipment. Thus, it was "completely natural to place at the disposal of the Chief of the General Staff as well that organ, through which the necessary direction can be given to the matter of aviation and through which the corresponding training of the officer personnel—pilots and observers in planes—can be carried out, as well as all necessary tests of flying apparatuses and their technical equipment conducted" (85). These considerations were supplemented by Sukhomlinov himself in his memorandum of November 19, 1911 to the Emperor.[8] As a sign of the importance thus now being accorded to aviation, he expounded at length upon the existing state of aeronautics and aviation in the army and the technical and organizational measures being taken for its development. It was in this context that he broached the issue of the impending transfer of aviation to the jurisdiction of the General Staff. Acknowledging the current concentration of aeronautical affairs in the Main Engineering Administration, he noted that such an arrangement was

> evidently not appropriate, since as a result of the specialization of the organs of the engineering department, the technical side of military aeronautics advances to first place. However, the ultimate goal of it [military aeronautics] is the service of the armies, the facilitating of their combat operations by giving them the most perfected means of reconnaissance and communication at present. Depending on this ultimate goal the general supervision both of training in the schools for military pilots and service in the aeronautical units should be united in the hands of that central organ which supervises in general the combat preparation of the army and which is in charge of reconnaissance and communications in the armies, that is the Main Administration of the General Staff.[9]

Sukhomlinov noted that this in itself should not exclude the participation of special technical bodies in the resolving of aeronautical questions. He then concluded his memorandum by summarizing his proposals; these general proposals were subsequently approved by the tsar.

Gen. Shishkevich himself, the newly designated commander of the Aeronautics Unit, also touched upon these issues shortly after assuming his new duties. In a draft copy of a report intended for GUGSh in October 1912, he commented at length upon the regulations that established and governed the Aeronautics Unit.[10] While claiming complete agreement in general with the proposed regulations, he nevertheless felt it necessary to propose appropriate modifications. The guiding principle of his comments was to specify in greater detail and extend the already broad authority of the Commander of the Aeronautics Unit in the overall direction of matters of aviation and aeronautics in the army. For example, his redaction of the responsibilities of the first section of the Aeronautics Unit read thus:

> the organization of aeronautical units and specialized aeronautical institutions; the working out of all questions concerning service and special training of aeronautical units and specialized aeronautical institutions; the study of the aeronautical forces and means of foreign states, and also aeronautical organization in Russia; the working out of rules for the free use of the air and the definition of zones of flight; the mustering of aeronautical units for summer exercises and maneuvers; the examination of questions concerning the staffing of specialized aeronautical institutions, and also the dispatch of personnel for instruction in aeronautical matters.[11]

This would give Shishkevich not only broad powers in regard to issues of training, staffing, and technology in matters of military aviation, but input as well into civil aviation in areas where it intersected with military concerns.

Shishkevich, like Sukhomlinov, also had his eye on the future. In regard to the jurisdiction of the commander of the Aeronautics Unit over cadres, he proposed adding an article which stipulated his authority "to select and present to the War Minister persons for the posts of commanders of the Officers' Aeronautics and aviation schools, and also special aeronautical institutions," as well as to decide upon the dispatch of officers of the OVS and aeronautical units to serve upon purchasing commissions and to conduct observation of production in factories. As at that time there existed only two aviation schools, such proposals were obviously made in light of the future expansion of the aviation services envisioned by all. Likewise, while commenting upon another article of the statute, Shishkevich noted it was necessary to include phrasing to specify that the Commander of the Aeronautics Unit was at the same time the commander of aviation schools and special aeronautical institutions: "serving as the motive for this inclusion is that in the plan of organization of aviation in the army approved by the War Minister, there is envisaged the creation of several schools and institutions with depots, whose activities will require unification."[12] Thus, in a service that was slated to undergo major expansions in the coming years, Shishkevich planned to institutionalize his authority from the start and expand the scope of his powers as the aviation institutions themselves grew.

These memorandums and reports addressed numerous issues regarding the expansion of the War Ministry's efforts in the field of aeronautics and aviation in general and the assumption of a prominent role in such efforts by the General Staff in particular. At one level were the basic criticisms raised by Dukhanin in his *Razvedchik* piece: the perceived slow pace of progress in aviation development, as well as a recognition of the problem of several organizations and institutions—the Department of the Air Force, the Aeronautics Section of the Engineering Administration, the Officers' Aeronautics School, the aeroclubs, etc.—often working without coordination or even at cross purposes. Although there is no evidence that this played a direct role in decision making, Dukhanin was not alone in his criticism. Similar concerns, though not always in such bald form, were echoed in aviation enthusiast journals and the popular press. The newspapers *Novoe vremia* and *Vechernee vremia* were to become particular nemeses of the military aviation establishment, often complaining of divided and dispersed effort; typical was an editorial in the May 3, 1912, edition of *Novoe vremia*, which complained that "the War Department is inclined to go its own way, and the OVF— its own..."[13] Criticism of a different sort was provided by N. E. Popov, the civilian pioneer who had originally trained the first Russian officers to fly in May 1910 until a crash ended his flying career.[14] In a 1912 brochure devoted to the achievements of Russian military aviation, Popov trenchantly observed that "it is necessary to immediately take away airplanes from the engineering department. One cannot subordinate eagles to moles!"[15]

The military press too appealed to the urgent demands of progress. Again, as a regular feature in the War Ministry journal *Voennyi sbornik* there were short reports detailing the rapid aviation developments abroad and exhorting Russia to emulate these successes, while *Razvedchik*, as we have seen, at times voiced its impatience even more directly. Dukhanin's and articles like his generally complained in some form about inertia and bungling in the development of aviation and cited the leadership of the military establishment for blame. Of course, in this brave new world of freedoms, however limited, in post-1905 Russia, such carping could be construed as unpatriotic slander or informed patriotic criticism, depending upon one's point of view. It was telling in and of itself that War Ministry policies were being debated in the public forum of the press with such vigor, with military aviation in particular attracting a wide degree of attention.

Equally significant in the transfer of aviation from the jurisdiction of the

Engineering Administration were the compelling signs that airplanes were indeed reaching the stage of practical military application. Again echoing the assertions of Dukhanin, when aviation had largely been a sideshow, the province of a handful of enthusiasts, haphazard organization had not been a problem; however, as the airplane increasingly developed into a serious piece of military technology, more careful attention had to be paid to it. For example, one of the more compelling signs of the figurative arrival of aviation was the successful participation of a squadron of airplanes from the Sevastopol Aviation School in the maneuvers of the Kiev and Warsaw military districts in the fall of 1911. The performance reviews were impressive. A telegram of the commander of the Warsaw military district, General-Adjutant Skalon, to the Grand Duke thanked him for giving his forces the opportunity to learn much, "to admire the courageous, skillful flights, to become convinced of the firm establishment of this affair in our native army and that enormous benefit which a brave pilot can be of."[16]

Such progress impressed an even more important figure: War Minister Sukhomlinov was instrumental in deciding that aviation would play an important part in the future of the Russian military. Sukhomlinov was apparently at first an aviation skeptic. In his memoirs, Grand Duke Alexander Mikhailovich noted that when he first told Sukhomlinov of his plans for the Department of the Force, Sukhomlinov between "paroxysms of laughter" asked if he were really serious about using such "playthings" for the army.[17] However, it was in response to a June 1911 General Staff report concerning the participation of aviation squadrons in maneuvers that Sukhomlinov had issued his peremptory decree to transfer aviation to the General Staff.[18] Like the experience of the Grand Duke himself learning of Bleriot's channel flight, it was a visit to the Sevastopol Aviation School in October 1911 that apparently left Sukhomlinov further impressed with the status and potential of heavier-than-air flight.[19] However, to fully bear fruit in technological advances or successful military applications, it was recognized that the activity of the aviation schools had to be buttressed by a framework of sound organization. Sukhomlinov's visit thus cemented his belief that the development of aviation must be pursued with urgency and determination. As detailed in his memorandum to the Emperor, the centerpiece of these new efforts was an ambitious plan for a large and rapid expansion of the aviation service, in terms of equipment, units, and the personnel. This decision itself meant that a rational, comprehensive direction of such efforts was now even more vital. Thus, the development of the aviation was beginning to produce the kinds of tangible results that could exert influence upon broader policy and galvanize the War Ministry into greater action in this area.

However, all of these considerations do not in themselves explain why aviation was specifically transferred to the direction of the General Staff. Instead, the creation of the Aeronautics Unit must also be placed within the broader context of the efforts of the General Staff to consolidate and expand its role as the pre-eminent institution in the Imperial army, charged with the most important tasks and attracting the best and brightest officers. In the historiography of the late Imperial army it has been noted many times that during his tenure Sukhomlinov intentionally sought to arrest this process and keep the power of the General Staff in check, both by limiting its formal powers and by appointing non-entities to the post of Chief of the General Staff. However, it is true nevertheless that once he was determined to pursue the development of military aviation, Sukhomlinov chose the General Staff for a leadership role as a prerequisite for serious planning and a comprehensive approach to the matter. To the rest of the army and outside observers—the creation of the Aeronautics Unit was commented upon throughout the civilian and military press—the transfer of aviation to the General Staff's jurisdiction was an unmistakable sign of the new attention and importance being attached to aviation by the military leadership.

At a fundamental level the transfer was the outcome of competing manifestations of professionalism. Sukhomlinov admitted as much when he noted that due to its specialization the Main Engineering Administration would engage military aviation overwhelmingly from a technical point of view. This in itself would tend to limit its development. To ensure that the airplane fulfilled its full military potential, what was needed instead was an organization that by its training, function, and purpose approached issues from a broad perspective, with the ultimate goal of improving the preparedness of the army in all respects. Zhilinskii thus explicitly appealed to the unique position of the General Staff as the one institution that fulfilled these conditions. To buttress such claims, he cited the leading role played by the General Staff in the application of an earlier revolutionary piece of technology, the railroad. The military development of railroads and their integration into the army had been accomplished through the establishment of a special railroad section of GUGSh; the same model was to be pursued with regard to aviation and aeronautics.

Moreover, throughout the arguments advanced by Generals Zhilinskii, Sukhomlinov, and Shishkevich, the attention paid to the issues of training and personnel was meant to ensure that the development of military aviation would not be treated merely as a matter of technology; instead, cadres would a fundamental part of the equation. This was evidenced, for example, in Zhilinskii's rationale for placing the Officers Aeronautics School under the jurisdiction of the Aeronautics Unit, which took over control of the admissions process to the aviation schools from the Engineering Administration. As noted, Gen. Shishkevich too sought to arrogate to himself even greater authority to supervise and direct the growth of the aviation service through control of staffing and appointments.

Above all, the expansion and empowerment of the General Staff in Russia, as in Europe, was a deeper response to the exigencies of technological advancement and military modernization in the decades before World War I. The assumption of control over aviation by the General Staff was predicated on the understanding that modernization involved far more than the simple acquisition of the newest aeronautical equipment; this in itself could have been left to the Engineering Administration. Instead, the acquisition, utilization, and assimilation of technology had to be conceptualized and executed within the context of the increasing scope, complexity, and multi-faceted nature of modern war, and the purpose of the General Staff was to manage this very process. Not surprisingly, General Shishkevich, the commander of the new Aeronautics Unit and himself a General Staff officer, supported the transfer; Colonel Odintsov, also an officer of the General Staff, likewise advocated that the General Staff assume a prominent role in the development of military aviation. These arguments found another source of support: line officers of the aviation services themselves, who made their desires known. Again, various articles, particularly those in the military press, purported to speak for such officers and criticized the existing status and organization of aeronautical affairs. A more direct statement was made when several officer-pilots in the first classes graduating from the Sevastopol Aviation School voted with their feet and refused to take postings in the newly forming aviation squadrons, despite the pressing need for trained pilots. Although it had been understood that officers agreeing to pilot training at the Sevastopol School would take these posts, due to the youth and institutional fragmentation of the aviation service, it seems there was no enforceable regulation obligating them to do so.

The ostensible reason for these officers' refusal, according to none other than Colonel Odintsov, was a lack of confidence in the Main Engineering Administration and its direction of aviation. In his personal letter to Zhilinskii of November 22, 1911, Odintsov noted that in the course of a conversation during Sukhomlinov's visit to Sevastopol, the War Minister had expressed his chagrin that these non-engineering officers had declined to join the aviation squadrons. He explained that "I considered it my duty" to report to the War Minister "those reasons which motivated the aforementioned officers to refuse to accept" the proposal of the Main Engineering

Administration. Odintsov listed these reasons as the following:

> 1. the deep certainty of all in the impossibility of successfully seeing this matter through to the end if aviation remains in the hands of the engineering department;
> 2. officers not of the engineering services are invited only for two years, after which they must undergo a special examination for transfer to the engineering services, which does not suit many;
> 3. the fear of ending up as stepsons of the engineering department;
> 4. the paltry increase in pay—660 rubles; and
> 5. with the exception of Kiev, the very bad stations.[20]

Odintsov however, added a caveat: "*but I am firmly convinced* that if aviation is separated from the Engineering Administration, there will not be one officer, *out of those who have finished*, who will under those conditions refuse service in the aviation squadrons."[21]

We do not know exactly how many officers refused to serve in aviation squadrons due to the perceived bias of the Engineering Administration, but there is little doubt that this letter reflected Col. Odintsov's own views on the matter as much as those of such officers. There is no evidence that this boycott had a direct impact upon the decision to transfer aviation to the auspices of the General Staff, while Odintsov himself would soon be at odds not only with the War Ministry but with the leadership of the Department of the Air Force itself due to his tendency to offer his unvarnished opinions, usually without being asked. However, this episode was indeed noticed and not appreciated at the higher levels of the War Ministry. In his memorandum to the tsar, Sukhomlinov advocated continued reliance upon the Sevastopol Aviation School of the OVF. However, he observed that in light of the significant sums being spent by the War Ministry—a 100,000 ruble yearly subsidy to the school—"the War Ministry is forced to categorically insist upon the obligation of officers completing the school to serve a definite period of time in the aviation squadrons, as of those pilots who finished the school in the present year, several did not wish to transfer to these squadrons, despite the urgent need for them."[22] Subsequently a two year term of service in the aviation squadrons was officially mandated for graduating pilots of both the Sevastopol Aviation School and the Officers' Aeronautics School. Thus, the creation of the Aeronautics Unit can be seen at least in part as an effort to master the issue of cadres in a broad sense.

The various motivations concerning the establishment of the Aeronautics Unit of GUGSh Staff indicate the extent to which the growth of the aviation service was creating an entire new complex of demands, expectations, and requirements. However, there was simultaneously a lack of agreement over exactly what these new expectations and requirements were to entail. It was the express task of the Aeronautics Unit to impose the requisite harmony, agreement, and cooperation among all involved. Yet, despite the urgency and high hopes that surrounded its establishment, the Aeronautics Unit found it difficult to provide the singular will and unified, coherent direction that was its mandate and which was seemingly so successful abroad. Instead, the work of the Aeronautics Unit was to be marked by a palpable taste of frustration and unrealized expectations. If some might complain that irony is an overworked trope in historical scholarship, in this case it seems entirely appropriate. The Aeronautics Unit encountered troubles from its inception.

The reasons were varied. Part of the problem lay with familiar nemeses of the Imperial army: bureaucracy and inertia. The dating of the documents quoted above illustrates the very sluggishness of the process of putting ideas into practice. War Minister Sukhomlinov had ordered in the summer of 1911 that the transfer of aviation to the General Staff be implemented as soon as possible. The Aeronautics Unit, however, only officially came into existence with War Department Order No. 397 in June 1912. Even then, the new institution did not truly begin its activity until September,[23] the delay being caused in part by the late arrival of its newly appointed commander, Major General Shishkevich. This was in turn caused by the unwillingness of his previous command, the staff of the Moscow military district, to release him to his new post. Shishkevich was eventually forced to appeal to GUGSh;[24] ina letter he explained that the local command would not let him go due to his role in the preparations for the jubilee honoring the Romanov dynasty, and was even planning an appeal to the War Minister to keep him there longer. He thus expressed the desirability of beginning his new duties as soon as possible, especially as the Main Engineering Administration had already ceased functions relating to affairs of aviation and aeronautics. The General Staff duly dispatched a letter to the Chief of Staff of the Moscow military district pleading for the release of Shishkevich. This was done, but it was not until early September that Shishkevich was finally able to report that the Aeronautics Unit had officially been activated.[25] Although it seems slightly comical, if nothing else the entire episode demonstrated that not all had an equal appreciation of the budding importance of aviation.

More importantly, as symbolic as the creation of the Aeronautics Unit of the General Staff was, it was not matched with equal substance. The entire staff of the Unit was comprised of only a handful of officers and clerks. This small complement of personnel was in part a reflection of the function of this institution. Its mission was the direction of all matters of aviation and aeronautics in the army, but this was not quite the same as complete control. In fact, the fundamental problem lay in the fact that by the very nature of its role—coordination and overall guidance—the Aeronautics Unit did little torationalize the existing organization of aviation in the army. Notably, one of the primary constituent units it was supposed to supervise—the Sevastopol Aviation School—remained officially autonomous of the War Ministry, as its parent organization, the Department of the Air Force, retained its quasi-public, largely independent status. The mandate of the Aeronautics Unit did allow it to give orders and directives to the Sevastopol Aviation School regarding issues of staffing and procedure, while the Grand Duke and his officers all pledged greatest cooperation to this new institution. In practice, however, as revealed in often testy correspondence, these were two different institutions which had their own, often conflicting, agendas.

The same could be said, for that matter, of the Aeronautics Unit and the Officers Aeronautics School. Although it had been transferred to the control of the Aeronautics Unit, the school also continued its work with largely the same structure, organization, and staff as previously. As noted, numerous officers of the staff had been deeply involved with aviation from its very inception and were experienced, knowledgeable pilots. Further, the status of aviation at the school had received an institutional boost with the long-planned transformation of the Aviation Section from provisional to permanent status in September 1912, an event contemporaneous with the establishment of the Aeronautics Unit itself.[26] In comparison, General Shishkevich had seemingly been selected for his new post in part because he was an aviation enthusiast and member of the Moscow Society of Aeronautics (though not a pilot), but also in part because he fit the rank and service requirements of the position. In such a young branch of arms as aviation, a combination of real experience, proficiency, and seniority was difficult to find in one person. The commander of the Aeronautics Unit was thus charged with giving orders to an institution whose experience, competence, and knowledge in both practical and theoretical terms exceeded his own. Not surprisingly, there were thus problems in this relationship too.

As this implies, the problem involved not just matters of organization and administrative control, but deeper issues of orientation, priorities, and authority relative to the development of military aviation. As illustrated by the establishment of the two sections (personnel and technical) of the Aeronautics Unit,[27] the War Ministry was attempting to unite and coordinate different functions that had been de facto assumed by different organizations. The Main Engineering Administration and to an extent the Officers' Aeronautics School emphasized experimentation, technical development, and were staffed primarily by officers trained as engineers. The Department of the Air Force and its Sevastopol Aviation School had placed their emphasis upon the intensive training of cadres and the practical application of military aviation. As stated explicitly in its charter, the goal of the Department of the Air Force was to build a powerful air fleet and train its pilots as rapidly as possible. Thus, in terms of both its orientation and policies—sponsoring the first group of officers to train in France, procuring the services of the civilian Efimov, recruiting officers from all branches of service—the OVF from its beginning directed its efforts more broadly than did the Engineering Administration or the Officers' Aeronautics School. If the Department of the Air Force and its Sevastopol Aviation School thus had somewhat of the atmosphere of a crusade, the Officers' Aeronautics School had that of a military university, while the Aeronautics Unit largely had the orientation of the administrative chancellery it was. Because such differences involved fundamental conceptions of mission, method, and goals, they were not so easily reconciled. This was further complicated by the injection into the equation of the General Staff, which had its own biases, policy preferences, and professional culture quite irrespective of aviation. Harmony between these institutions would thus prove elusive.

Even before the official establishment of the Aeronautics Unit, conflicting motives, desires, and attitudes were already in the forefront. These could serve to muddy the General Staff's commitment to aviation even as it was asserting its authority in a manner bordering on arrogance. Vasilii Rebikov, an engineer and member of the Department of the Air Force, sent a revealing letter to the Grand Duke Alexander Mikhailovich in June 1912, in which he reported the results of his separate meetings with Zhilinskii and General Shishkevich.[28] Rebikov's second meeting with Zhilinskii concerned the relationship of the Department of the Air Force and GUGSh in view of the impending transfer of aviation to the latter. Rebikov informed the Grand Duke that he would try and reproduce Zhilinskii's words as accurately as possible; if he did so, it is clear that the Chief of the General Staff was not shy about expressing some home truths. Zhilinskii began by stating that

> I did not at all insist that aviation be under the General Staff; I have enough work without it. I am interested merely in results of aviation useful to us, and am not so familiar with it to order in what way to achieve these results. His Highness [the Grand Duke] wished that aviation be in the War Ministry and expressed this several times, and the War Minister, my commander, considered it most convenient and correct to entrust its administration to me.

Zhilinskii went on to note that the higher organization of aviation in the army was already approved and to change it now would be impossible; he added for emphasis that "I personally will not even lift a finger to alter it."

Addressing the status of the OVF, he observed that "the Air Force is a free institution, a sort of Volunteer Fleet, and is useful to us as a school." He noted that the Sevastopol School and its affairs had been wonderfully organized and managed by Colonel Odintsov,

> but in the matter of the organization of aviation service he is of no use at all, as he indulges in fantasy and in general behaved irritably and extremely tactlessly, telegraphing the War Minister several times carrying on agitation. The Grand Duke behaved extremely restrained and calmly the entire time and by this demonstrated his strength of character and impartiality. The knowledge of His Imperial Highness in the matter of aviation is great, we all know this and we depend on his help. We have in the General Staff a Special Committee, of which I am the chairman and a member of which by the way is the Grand Duke Sergei Mikhailovich. I will appeal to the Grand Duke Aleksandr Mikhailovich with a request to help us in this Committee and am counting on his agreement.

Zhilinskii therefore concluded that

> thus, I consider the activity of the Air Force beneficial and useful to us and in what I can I will help it; if in the future a closer unity turns out to be necessary for some reason, then only in this case will we try to find a compromise and reach agreement with the Grand Duke and, slowly, not rushing, and deliberately take the necessary steps at all levels. Thus, the Air Force is useful to us as an independent Establishment; I do not consider it an 'Institution'; I hope that after this exchange of thoughts you understand me and your activity will be useful to us.

Although this incident was perhaps as much evidence of Zhilinskii's questionable commitment to his position—the reluctant aviation administrator would later in his career beg out of remaining Chief of the General Staff— it said a great deal about the attitude of the General Staff as a whole towards military aviation. Zhilinskii expressed both a commitment to the needs and priorities of the army as a whole as well as a keen awareness of the institutional importance of the General Staff. General Shishkevich's own comments in his reports concerning the proper role and powers of his new position paralleled such thinking. The General Staff clearly considered its role in the development of aviation as one of supervisor, not equal.

The Plan of Organization of Aeronautics and Aviation

It was in this context that the Aeronautics Unit turned its attention to its primary task: the elaboration of a comprehensive plan of organization of aeronautics and aviation in the army. Begun first by the Main Engineering Administration and then taken up by the Aeronautics Unit, this was an blueprint that covered all areas of aeronautics and aviation in the army—from cadres, service, and schools to the material composition and technical supply of units—with its goal the successful management of future growth. By necessity, the plan thus confronted the fundamental questions and challenges emphasized in the projects proposed by the Department of the Air Force and General Kovan'ko. In particular, the issues of industrial resources and trained cadres often took center stage.

This project, however, underwent numerous redactions throughout the three-year course of its development, reflecting the buffeting winds of financial constraints, political considerations, changing technology, and rising expectations.[29] Further delays were caused by the General Staff's own decision to temporarily suspend all organizational measures while Gen. Shishkevich acquainted himself with the status of aviation abroad.[30] To an extent symbolic of the entire course of Russian Imperial aviation, while some of its provisions were to be implemented piecemeal, the plan as a whole, including some of its major programs, was to receive final confirmation only in time to be embodied in the Large Program of 1914. The latter, alas, was negated by the outbreak of World War I.

Nevertheless, in his memorandum of November 19, 1911 to the Emperor, War Minister Sukhomlinov emphasized several fundamental tenets, derived from the "rich experience" both domestically and abroad, that would guide aeronautics and aviation policy.[31] Foremost among these were the now clear advantages of airplanes in comparison to balloons. While not totally rejecting balloons, dirigibles, and related equipment, "the War Department should now direct all of its efforts to the soonest possible supplying of the army with airplanes." Regarding the tension between buying abroad versus building at home, Sukhomlinov noted that "with the goal of development and encouragement of industry" that produced aeronautical equipment, primarily airplanes, "the War Department should be given the right to order these apparatuses primarily in domestic factories, even if at more expensive prices, than abroad" (139). He added that it went without saying that this would be done under the condition that these factories be obligated to supply apparatuses of the latest design and most perfected construction. Nonetheless, Sukhomlinov clearly recognized in particular the need to develop a domestic aviation industry in the interests of military necessity and national security. Sukhomlinov emphasized that the primary goal of the War Department was the soonest possible formation of eighteen aviation squadrons, consisting of six planes each, for which funds were now being requested from legislative institutions. He ideally foresaw the eventual creation of up to 45 aviation squadrons (12 planes each, half of them reserve) for the field army, to serve the army corps as independent arms of reconnaissance and communication, and other units for the fortresses. He further observed that the Sevastopol aviation school should continue its existence, as it was exactly in the coming 3–4 years that the School, along with the Officers' Aeronautics School, would need to expand its efforts to meet the need for cadres for the new aviation squadrons. He proposed indeed to expand admissions both to the Sevastopol School and the OVS and perhaps to open several separate branches of each (140).

With these "fundamental" principles in mind, it was up to the Aeronautics Unit to work out the particulars. In a report of October 1912, co-signed by General Shishkevich, Chief of the General Staff Zhilinskii presented Sukhomlinov a detailed reworking of the draft plan of organization itself and proposals for the reorganization of military aeronautical service in general.[32] The first subject broached was the training of personnel for the new aviation squadrons. A certain urgency regarding the establishment of the military's own schools was again a primary theme: the report emphasized the need for the soonest possible establishment of aviation schools of the War Department, "completely autonomous and independent of the existing Officers Aeronautics School" (136). The plan subsequently proposed three schools, two of them in European Russia and one in Tomsk, Siberia. The report acknowledged that the selection of these locations had been governed by climactic conditions and considerations of the least possible expenditures of traveling money for officers assigned to the schools, as well as the opportunity to begin operations quickly. With the latter consideration in mind, the first school would be established by transforming the existing Aviation Section of the OVS and its personnel and resources (136–37).

In the meantime, the report recognized the necessity of the continued use of civilian aviation schools, "as a temporary measure" (a qualifier which moved Sukhomlinov to underline it and write in the margin "precisely"). Yet the report proposed working out a temporary regulation by which a predetermined number of officers would be trained on aircraft systems used by the War Ministry. For each officer who successfully passed an examination program for the designation of military pilot, worked out by the war department, the school would be paid 500-700 rubles. This system would replace the present practice of lump yearly subsidies to the Imperial All-Russian Aeroclub and other such organizations (136). This was indeed the policy the policy that was subsequently implemented. In general, Aeronautics Unit reports and memoranda frequently expressed the desire to free the War Ministry as soon as possible from any dependence at all upon civilian aviation organizations, and their schools in particular.

The report further proposed the establishment and organization of aviation stations, which would serve as bases of concentration for united groups of aviation squadrons, now to be separated from the aeronautical companies (138). These stations would serve as centers for testing, training, repair, maintenance, and supply in peacetime; in wartime they would function as bases for the dispatch of squadrons to the appropriate army corps and for the maintenance of equipment. They were to have a permanent administrative staff which would oversee the equipping of the squadrons, the supervision of the aerodrome, and also direct the training of the squadrons stationed there. There were to be four in European Russia, three in Asia. Counting the smaller groupings of squadrons in the Caucasus, Turkestan, and Siberia, in all this was to provide the army with 27 corps aviation squadrons, 10 army squadrons, and 5 squadrons of special assignment, providing for a total of 42 squadrons, which did not include 8 fortress squadrons; this would necessitate adding two more aviation bases to ensure that each had a maximum of seven. In an explanatory footnote, it was further proposed that in time of war the stations could serve as points for refresher training of soldiers in the category of pilots called out of reserve, and also as places for the formation of volunteer aviation squadrons (138–40). The report noted again that the specific statutes and regulations governing the centers were being presently worked out by the Aeronautics Unit.

The report also addressed issues of staffing and personnel, citing deficiencies in the existing organization of the aviation squadrons that had been worked out by the Main Engineering Administration. It noted that the regulations provided for four officer-pilots assigned to each squadron, one of whom simultaneously functioned as squadron commander, but that such a system "does not meet actual requirements." The report argued instead that in light of the administrative duties of the commander, particularly regarding the oversight of the squadron's significant component of materiel, it was impossible for him to be a pilot as well. In addition, there had to be a permanent presence to maintain communication with corps and army staffs and direct squadron operations accordingly. If the commander was simultaneously a pilot, "there cannot be any talk of direction of the activity of the unit, on the contrary it may bear a random and perhaps even undesirable character." In light of this the report felt it was necessary to add at least one officer-pilot to the complement of a squadron and separate the squadron commander from the ranks of active pilots (141).

In a later section entitled "On Service in the Aeronautical Units" the report addressed the compelling issue of cadres in more detail, discussing both tangible issues of terms of service and promotion but also a broader ethos of purpose and duty in the new squadrons. In the opinion of the General Staff, this was a matter that deserved as much or more attention than the technical and administrative refinements described above:

> one of the important aspects of a dependable organization of the matter of the aeronautical defense of the state is the provision of the aeronautical units with good personnel; for this it is necessary to create conditions of service in these units which not only provide for materially but also provide the hope of moving along the steps of the service hierarchy further than the rank of lieutenant colonel, as takes place now.[33]

The report explained that under the existing organization, officers of the aeronautical units, with the exception of the aviation squadrons, were listed in the engineering services, but did not have the right of transfer to sapper battalions. Thus, having achieved the post of company commander, with the rank of lieutenant colonel, they lost hope of further advancement, as in the entire aeronautical service there was only one post that carried the rank of colonel; "without any doubt the thought of the cessation of a further career has an effect on the activity of a company commander of an aeronautical company, weakening his energy." The report continued that

> on the other hand, the existing regulation on service in the aviation squadrons, providing temporary material support during one's stay in these squadrons, does not contain elements of service movement and the possibility of the continuation of service in the aeronautical services [*voiskakh*]. Meanwhile, with the development of military aeronautics at the present time and the imparting of an active character to it, the need for an excellent staff of officers has significantly grown, and not only for service in the aviation squadrons, but even in the aeronautical services, which should serve as the source of additional personnel for the presently developing aeronautical institutions (aviation schools, centers, etc.).[34]

In light of this the report recommended several necessary measures. It proposed first to remove the entire officer cadre of the aeronautical units from the staff of the engineering services and transfer it to the jurisdiction of the Chief of the General Staff. It then proposed to form a special aeronautical corps for this officer cadre, which would provide the opportunity for service advancement in the corps itself without transfer to any other units. Lastly, depending on the foregoing, it proposed to establish a new procedure for staffing the aeronautical units with officers. Addressing these proposals in greater detail, the report noted that with the establishment of the three proposed aviation schools, the retention of the present aeronautical school, the establishment of the nine aviation stations and companies and the reorganization of the aeronautical units, the proposed aeronautical corps would include 4 posts for the rank of general, 15–18 for the rank of colonel, up to 30 for lieutenant colonels, about 80 for captains and around 350 for *ober* officers. Thus, "in this way this corps will contain in itself elements of service advancement for the best officers and consequently there will exist some incentives for effort and the maintenance of energy."

The report then explained the new procedure for staffing the aviation units. This procedure, like the proposal of Colonel Odintsov for a special aviation *uchilishche*, was an attempt to deal with the fundamental problem of establishing an institutionalized pool from which to draw trained, competent, and committed cadres. The General Staff's response was basically to recognize the need to establish aviation and aeronautics as a separate branch of service. The report thus recommended that the basis for entry into the aeronautical services should be service in the aviation squadrons. These squadrons would be staffed with officers from all branches of service without exception, but only those who satisfied specified conditions, who completed the full course of an aviation school, and who passed the examination for the rank of military pilot. Each of these officers would then spend three years serving in the aviation squadrons, in the course of which they would be listed by branch of service in the service lists of their units, but not counted as filling a vacancy. Only for reasons of health, hindering him from flying, could an officer return to his unit before his three year stint in the squadron ended. Upon completion of the three years an officer would be granted the choice to return to his unit or to continue in the aeronautical services.

The report recommended that officers wishing to continue in the aeronautical services should then spend a year at the Officers' Aeronautics School, which should be transformed along the model of the officers' artillery school. Those officers successfully completing the year of instruction would then be listed in the aeronautical services and would receive assignments in the aeronautical institutions, companies, and squadrons, in the latter as commanders, not pilots. Those who did not finish would be returned to their units. Finally, in order to maintain the officer complement of the aeronautical units at the proper level, the report recommended retaining the existing age limits for service in these units: 35 years old was the maximum age for pilots, 50 years old was the maximum for those serving in guided balloons, and 55 years old was the maximum for those serving on anchored balloons. Lastly, the report asked the War Minister to approve the considerations laid out in the report for further reworking of the measures by the Aeronautics Unit. Sukhomlinov duly approved.

Six months later Gen. Shishkevich was to provide a secret report to Zhilinskii regarding the development of aviation and aeronautical affairs in the army.[35] Far from a mere status report, it was as much a marketing tract trumpeting the accomplishments of the Aeronautics Unit in the scant eight months of its existence. However, it also provided a concise, if not unbiased, historical overview of the development of aviation policy in the army and thus shed further light on the General Staff's own perspectives in this area. The report began by noting that already in 1909, following the first successful flights of Wilbur Wright and the Reims Aviation Week in France, the Russian War Department, through the Main Engineering Administration,

> had turned its attention to the arising matter of aviation in general and to the application of it to military affairs in particular. But owing to the small familiarity with this new means of combat in military circles not only among us, but abroad too, the first steps in this direction could not have a sufficiently systematic character and besides, owing to the absence of funds at the disposal of the War Department for the establishment of the new and very expensive materiel there could be no talk of any kind of the correct organization of the aviation services.[36]

The report summarized the subsequent efforts of the Main Engineering Administration, including its plan of organization of aviation service and the establishment of the Aviation Section under the existing Officers' Aeronautics School. On April 7, 1911, a bill had been presented for examination to the State Duma for the release of credits to fund the new Aviation Section. Shishkevich noted, however, that "despite the large interest shown by the State Duma in the organization of the matter of aeronautics in the army, as expressed in its frequent inquiries in this sphere, it could not find time, however, for examination of the bill submitted to it in the course of the entire year," and it was only in the spring session of the following year that the bill was taken up, receiving the force of law only on May 10, 1912. Shishkevich observed that the bill providing credits for the formation and support of the 18 aviation squadrons received force of law at the same time, although it had been submitted to the Duma a half year later.

In light of the above, the report noted that the middle of 1912 had to be considered the beginning of the organization of aviation service in the army. Yet it detailed the modest status of affairs that GUGSh had inherited when it assumed jurisdiction: only eight aviation squadrons could be considered to exist; only five of them had the regulation complement (four) of officer-pilots, while for the remaining three squadrons there were all of six officer-pilots. In addition to these 26 pilots there was only one pilot from the ranks. It was the same story for equipment: only 4 squadrons had the full regulation complement of planes (6), while for three others there were all of eight planes, and one squadron had no planes at all. This gave a total of only 32 planes on hand. Lastly, regarding spare parts, motors, and supplies, the squadrons were far from fully equipped as well. The report continued by proudly describing the subsequent energetic activity of GUGSh in this sphere. It noted that while it was still working out a complete, final plan of military aeronautics and aviation, the General Staff had "energetically" proceeded with the continued formation of the first 18 squadrons and the provision of them with the requisite materiel. The Aeronautics Unit could thus report that after eight months, the 18 squadrons had been completely formed and had their full complement of 72 officer-pilots, and in addition 10 pilots of the lower ranks. In addition, officer-pilots above the official complement had been assigned to all squadrons, so that in all there were 105 pilots in the squadrons. When those officers who had been sent back to their units were added, the figure became 126 pilots in all.

This raised the issue of surplus pilots, with the report noting that "it could be held against" the War Department that not all trained pilots were being utilized in the squadrons and that some had been sent back to their line units. However, it added that in light of the procedure for handling organizational questions this was entirely normal: receiving the necessary legislative approval for the requisite credits to expand the number of units and officer complement took time, as did funding the new construction for the squadrons. The report also explained the surplus of pilots by citing the

> intensive training of them; in light of past alarming political events, pilots are being trained not only in the school of the War Department but in the Sevastopol Aviation School and in the private aviation schools of the Imperial Moscow Aeronautical Society, and this surplus does not trouble the War Department in the least, as it is a reserve in wartime and provides for the further formation [of squadrons] without delay.[37]

The report added that the fear that such pilots would forget how to fly was not a cause for concern, as practice had shown that only a small amount of practice was necessary for a trained pilot to regain his skills. It noted that in the past winter pilots trained on Farman aircraft had all been re-trained on Nieuport machines without difficulty, in light of the equipping of the new squadrons with Nieuports.

The report emphasized that in this way the Aeronautics Unit had spent 4,655,492 rubles in the eight months since questions of aeronautics had been transferred to it. Given the number of personnel provided for the Aeronautics Unit, "this serves as evidence of the enormous work of the staff of this unit." The report continued that the Aeronautics Unit, not limiting itself to the complete formation and equipping of the eighteen squadrons, had worked out a detailed plan of organization for aviation and aeronautics in the army. This plan, with some modifications, had a great deal in common with the draft presented to Sukhomlinov six months earlier. In all, the fulfillment of the plan would necessitate an expenditure of 44 million rubles, and General

Shishkevich noted that he understood that this was dependent upon the assignment of the necessary credits. The report noted that the War Department intended to implement the plan in the course of 1913, 1914, and 1915. If the State Duma approved the credits requested amongst other measures concerning aeronautics, another 21 aviation squadrons could be formed by the end of the current year. In this way, it concluded, in the course of one year the number of aviation squadrons would grow from 8 to 40, "and in this regard we will significantly surpass our western neighbors."[38]

This was the bright future for Russian military aviation envisaged by the leadership of the General Staff and the Aeronautics Unit. There was but one last piece of the puzzle to tend to. In addition to his work on the plan for aviation and aeronautics, Gen. Shishkevich already in March 1913 proposed a significant expansion in the size and staff of the Aeronautics Unit itself.[39] He noted that upon the formation of the Unit the size of its staff had been established temporarily for only one year. However, the explosive growth of the aviation services in the last year, measured alone in the volume of paperwork generated, had already rendered this staff insufficient and made expansion absolutely necessary. This was all the more true as this growth would accelerate further with the implementation of the measures contained in the plan of organization. For this reason Shishkevich proposed an expanded complement of personnel and credits to ensure that the Aeronautics Unit would be able to accomplish its valuable work.

Although it is usually just noted in passing in the small literature on the Imperial air forces, and receives just as little comment in the burgeoning literature on the General Staff in this period, the creation of the Aeronautics Unit was a significant event. As both an institutional and cultural response, in one sense it bore witness to broader, pan-European trends attendant to the challenges of military aviation. Given the rapid development of the airplane, all of the European militaries confronted and at times struggled with issues of organization, administration, and command. The options included among others placing the air services under the supervision of the engineering departments, the communications services, or the General Staff; parceling out aviation units to army corps and subordinate units; or establishing aviation as a de facto stand-alone service within the army. Respectively, the French created an independent Inspectorate of Aeronautics, the Germans subordinated aviation to the Commander of Railways and Transport, while the British initially placed aviation under the command of the Master-General of Ordinance, only to create an independent Directorate of Military Aeronautics in September 1913.[40] Nor was the Russian Empire alone in possessing an energetic public organization devoted to assisting the military in the national cause of aviation; the French Aeronautical League and the German Air Fleet League were both prominent in this regard.[41]

Yet the existing state of military aviation in Russia, as well as the response embodied in the creation of the Aeronautics Unit, illustrated issues of reform, institutional competition, and professionalism unique to the tsarist officer community. Again, the development of the Russian air services must be considered under the rubric of military reform in a broad sense, defined as an army's continuing process of self-examination, adaptation, and improvement. Historically, the hallmark of a successful reform is generally considered to be institutionalization, whereby reform ceases to be a matter of ideas and individual personalities and instead becomes one of policies and institutions. While this may not always entail the creation of entirely new organizational structures, it must entail some fundamental change in the military etsablishment's institutional culture. The creation of the Aeronautics Unit and the transfer of aeronautics and aviation to the jurisdiction of the General Staff indeed seemed to fit this model. It was a conscious attempt to put into place an institutional framework that would successfully direct diverse efforts to develop military aviation as was being done abroad. It was at the same time another step in the development and permanent establishment of a new branch of service and profession of arms. In contrast to the more gradual and long-term development that generally marked the rise of military as well as civilian professions, this process assumed a particularly concentrated and condensed form in the field of aviation. It involved the elaboration of criteria for recruitment, the setting of standards for technical education and training, the codification of regulations to govern aviation service, and the standardization of procedures to guide the selection of aviation technology and related equipment.

Moreover, the corps of General Staff officers was uniquely imbued with a sense of its own competence as well as the gravity of its responsibility as the one institution with a mandate to manage the challenges of modern warfare in the widest sense. It was thus precisely when the airplane was recognized as a technology with potential for broad application that it was considered both fit and necessary to be appropriated by the General Staff. Now vested with the imprimatur of this institution, the airplane was meant to assume a new importance within the efforts of the Russian Army to eradicate the stain of 1904–05 and prepare for the war many saw just over the horizon. The creation of the Aeronautics Unit was to ensure that the development of military aviation and aeronautics would be pursued with the best interests of the army as a whole in the forefront. This was a role that the Russian General Staff had been claiming for itself in over twenty years of self-advocacy.[42]

However, as noted, this claim had been badly compromised in the eyes of many regular officers by the performance of the General Staff in the Russo-Japanese War. The post-1905 period was thus in part marked by the efforts of the General Staff to reclaim its mantle of leadership and arrogate to itself the broad powers and latitude necessary to see that its goals were met. Although this was a definition of the function and power of the General Staff that was by now well-established in Germany, this was still a fairly hotly contested issue in the Russian army, with many suspicious of such a patently foreign model. The newly symbiotic relationship of the General Staff and aviation can thus also be seen as another attempt by the General Staff to consolidate its status as the source of intellectual and moral authority within the army. The creation of the new Aeronautics Unit not only asserted the growing importance of the airplane, it reemphasized the preeminence of the General Staff. The General Staff's faith in its unique ability to realize its vision came through its cultivation of a strictly professional ethos and culture, achieved through careful selection, rigorous education and training, and a well-developed sense of group identity. All of these elements are included in William Fuller's criteria of professionalism; more importantly, the General Staff satisfied the essence of Fuller's definition by maintaining a palpable professional *consciousness*.

Yet the professional ethos and culture of the General Staff was largely an exclusive ideal. The lack of rotation of General Staff officers to line commands, the contempt for those who failed to complete the third year, the resentment of line officers at the pretensions of staff officers, all testify to the fact that General Staff officers saw themselves as a vanguard of leadership, not equal comrades in a common cause.[43] The exclusivity of the General Staff project was recognized as such by many line officers, and was thus attacked by many as a perceived threat to the culture and traditions of the Russian army. The virtues and vices of this state of affairs manifested themselves in the sphere of aviation. Notably, Gen. Zhilinskii seemed to suggest that aviation was not yet even worth the General Staff's valuable time. However, with the growing significance of the airplane, the General Staff's predisposition to the mantle of leadership quickly asserted itself. That there were already existing institutions and individuals who considered themselves at least as knowledgeable and experienced in the field of aeronautics created tensions.

Notably, in regard to the engineering service, the General Staff confronted a corps of officers as professional as itself. This service also maintained high standards of qualifications, admission, and training, a strong esprit de corps, and a well-developed institutional culture. It possessed quality educational institutions, kept abreast of broader developments in the sphere of engineering and science, and by its very nature dealt frequently with questions of new technology. In fact, it was in stated recognition of this vigorous professional culture that the General Staff assumed control over aviation. By training and inclination, engineers would take a "too narrow" and technical view of aviation, thereby hindering its broader development. These were the professional opinions of the General Staff, but that does not mean they were accepted by those officers, largely trained as engineers, who staffed the Officers' Aeronautics School and its Aviation Section. They instead maintained their own conceptions and attitudes regarding the proper path of the development of aviation.

There existed too the unique institution of the Department of the Air Force and the Sevastopol Aviation School. While he could claim neither a General Staff diploma nor years of advanced engineering education, the Grand Duke could claim that he had done more thus far for Imperial aviation than anyone. Likewise, convinced of the urgency of their mission, the diverse group of officers of the Sevastopol school set to work with a will, creating a new and distinct tradition of aviation service and a legacy of success upon which to build in the future. While not denying the vibrant professional cultures of the engineers, aeronauts, and the General Staff, they could indeed question the relevance of these cultures to aviation. Mikhail Efimov was a civilian, but he knew how to unlock the secrets of the airplane; must he defer to generals, bureaucrats, or engineers, who possessed only rank or obsolete knowledge?

Thus, the place of the newly established Aeronautics Unit largely mirrored the status and role of the General Staff in the army as a whole. Beneath the overlay of its claims to leadership and its own motivations and perspectives regarding the challenge of the airplane, much beyond its closed ranks remained unchanged. Thus, instead of promoting coordination and unity of effort, the newly established administrative framework helped maintain and even propagate competing visions of the demands, requirements, and future path of military aviation within the Russian Imperial army.

5| Personality, Professionalism, and the Politics of Command

The various projects to develop Russian Imperial military aviation were comprehensive responses to the specific challenge of military modernization posed by the airplane. In these projects the commanding officers of the various aviation institutions actively sought to place their own stamp upon the structure, organization, and culture of the budding Russian air forces. These projects, however, dealt primarily with issues of broad policy and general principles; for a variety of reasons, moreover, a good many of their recommendations were to remain unrealized. An equally important forum for interpreting the demands of military aviation, however, was the daily exercise of command and administration that engaged the respective commanders of the aviation institutions and line units. The relevant issues were intricately bound up with the fundamental questions addressed in the projects: identifying goals and priorities, articulating institutional interests, effectively utilizing personnel, and exploiting resources.

As such, these issues elicited similar institutional differences in attitudes, priorities, and methods. However, within this sphere such differences were no longer confined to paper, but were instead played out within the larger context of the norms of hierarchy, deference, and authority that defined relations between officers. In the milieu of self-examination, criticism, and reform that prevailed in the post-1905 tsarist officer community, however, such norms were themselves being scrutinized, questioned, and debated. There were numerous attempts, both large and small, official and unsanctioned, amongst both various military elites and officers of the line, to redefine the terms and channels of authority within the tsarist army. Pertinent issues included emerging models of professionalism, the role of personality, the weight of tradition, and the lure of foreign models of service.

In light of such issues, for commanding officers within the aviation service, everyday issues of administration and management—exercising initiative, defining standards, training subordinates, building morale, combating inertia, expressing dissent—assumed added importance. Imbued with a faith in the revolutionary importance of the airplane, theirs was a sense not of hesitance and pessimism but of urgency and optimism. The freedom to pursue their cause was again augmented by the sheer novelty of aviation as a technology. Those in positions of authority thus had ample opportunity to attempt to impose their ideal visions of service and duty through their own example of conduct and leadership. Through memos and reports, letters and telegrams, these commanding officers appealed to disparate conceptions of authority, hierarchy, and competence, resulting in frequently contentious relationships among them. The politics of command in the aviation service thus provide further insight into the broader dynamics of modernization and reform within the Russian officer community as a whole.

Authority and Leadership

The numerous challenges and demands faced by the senior officers of the army and aviation service are by now familiar. As seen, none of the main actors in the aviation services were shy and retiring in expressing their opinions regarding these matters and the overall course of aviation development. The underlying tensions in their differing goals and approaches were often exacerbated in their personal exchanges. Confidence could be construed as insolence, initiative as presumptuousness, self-assertion as insubordination.

Colonel Odintsov, commander of the Sevastopol Aviation School of the Department of the Air Force, was most often the target of such complaints. In the critical assessments offered about him one can detect many of the characteristics of the negative stereotype of the "reforming officer" in the tsarist army. These included charges of opportunism, careerism, and self-aggrandizement; in short, a lack of respect for hierarchy, accepted routines, and official channels.[1] As we have seen, Odintsov was not averse to criticizing what he saw as inertia or incompetence on the part of the War Ministry, be it in the matter of jurisdiction over aviation or in the organization of staffing and training. What is equally clear is that Col. Odintsov was dedicated, energetic, and competent. Both memoirs and sources from the period reveal that he had support amongst much of the permanent and temporary staff of the Sevastopol School and others active in aviation circles; he was widely regarded as one of the leading figures in Russian aeronautics.[2] Even Chief of the General Staff Zhilinskii, who berated Odintsov and his pretensions in his conversation with the Grand Duke's representative, had to admit that he had done a wonderful job organizing the Sevastopol Aviation School and its training procedures.[3] It was Odintsov's ambitious attempts to have input into broader matters of policy, which he felt were integrally connected to the successful development of military aviation as a whole, that ultimately got him into trouble. It was this conviction and zeal that caused him to behave "irritably and extremely tactlessly" in his relations with senior officers. With the fervor and certitude of the new convert, Col. Odintsov could brook neither apathy, lethargy, nor ignorance among his fellow parishioners.

At times Odintsov was effective in his relentlessness. In a letter to the Grand Duke Alexander Mikhailovich in November 1911, he observed that his recent request for the assignment of twenty supplementary officers to the general class of the Sevastopol Aviation School had been denied by the Department of the Air Force.[4] He thus explained that "perhaps I did not sufficiently substantiate my report; therefore I will allow myself to again request" the assignment of said officers. Reiterating his reasons, he noted that if the school, "under comparatively unfavorable and difficult conditions," had been able to train 70 people to fly in the current year, then in 1912 it had the opportunity to train 140; he added that he did not foresee any extra costs arising from adding these extra officers, which was always an important consideration in military affairs in tsarist Russia. Odintsov concluded that

> the comparatively favorable results achieved by the school in this year, it seems to me, give me some right to expect that my petition can be treated with some confidence. As a result, if agreement to this assignment follows, I foresee for your Imperial Highness full moral satisfaction, that the school founded by you is so productively working apace. The OVF and all of its staff is proud of the fact that it will give Russia such a large number of pilots, and I, having turned them out, will thus receive the opportunity to thank Your Imperial Highness for giving me the opportunity to apply my powers and knowledge to the matter which interests me.[5]

Odintsov's appeal was apparently persuasive, as the Department of the Air Force subsequently petitioned the War Ministry for the dispatch of the extra twenty officers, a request which was granted, and in February 1912 further requested that the yearly officer contingent be permanently increased to 60 officers instead of the previous 40.[6]

More often, however, Col. Odintsov's unyielding temperament resulted in numerous clashes with officers both junior and senior. For example, responding to the mountain of criticism that accompanied the Imperial All-Russian Aeroclub's efforts in organizing the St. Petersburg–Moscow flight contest in 1911, the naval pilot Lieut. Iatsuk noted a "characteristic" report in the press that Col. Odintsov "lodged the complaint that the organizers did not seek his advice. What could Col. Odintsov have advised?"[7] Odintsov's report regarding the need to establish an aviation *uchilishche* was also, as were many of his contributions, apparently unsolicited. It was but one of the incidents that caused General Zhilinskii to deplore Odintsov's "fantasizing," meddling, and agitation with the War Minister.[8] Odintsov's personal letter to Zhilinskii regarding the refusal of newly trained pilots to serve under the Engineering Administration was proffered and received in the same vein.[9] In this letter Odintsov further felt compelled to add his criticism of the Main Engineering Administration's choice for a squadron station in Spass, Primorskii oblast. He noted that the "beloved" Farman aircraft chosen by the GIU to equip this squadron would be able to fly a mere 65 days a year due to climate conditions, adding that while Bleriots or Nieuports would be able to fly 100 or 150 days a year respectively, the GIU possessed none of the two latter types. He concluded that conditions for the Kars squadron also compelled careful consideration of the choice of aircraft type there, "but in any case it was not the Farman beloved by the Engineering Administration."[10]

Such friction between Col. Odintsov and the authorities of the War Department, a state of affairs hardly propitious for harmonious cooperation in the development of military aviation, eventually played a role in his ouster as commander of the Sevastopol Aviation School. This was a minor scandal that produced a variety of gossip and received a good deal of attention in the popular press. His downfall was associated with what was to have been the crowning achievement of the Department of the Air Force: the establishment of a new, permanent home for the Sevastopol Aviation School along the river Kach', north of Sevastopol. Spacious and well equipped, it was to be a marked improvement over the cramped environs of the old location within the city itself. However, the construction and equipping of the aerodrome ran into delays, financial wrangling, and second-guessing, for which Odintsov received the brunt of the blame.[11]

This unsettled state of affairs was exacerbated by a controversial decision, ridiculed in hindsight by historians and aviation officers themselves, to select modified French Nieuports as the aircraft of choice for the army's new aviation squadrons.[12] This selection came at the direct behest of the Grand Duke Alexander following successful training flights on these aircraft in June and July 1912 by two officers of the Sevastopol Aviation School, Lieut. Dybovskii and Staff-Capt. Andreadi. Dybovskii's flight, among the longest at that time, covered 2,235 versts along a route of Sevastopol–Kharkov–Moscow–St. Petersburg; Andreadi followed with a more impressive flight, covering 3000 versts along the route Sevastopol–Odessa–Kharkov–Moscow–St. Petersburg[13] Immensely impressed with these feats, the Grand Duke decided that the aircraft in question should become standard. Although such decisions were ostensibly the sphere of the General Staff, this choice subsequently became policy. At a time when there was still a great deal unknown about the flight capacities of various aircraft systems, and indeed when new designs and modifications were being introduced regularly, imperial prerogative once again made itself felt. Odintsov, who had already weighed in with his criticisms of the Engineering Administration's preferences, supported the selection of the Nieuports. However, there were apparently other concerned officers, both within and outside the Engineering Administration, who were neither pleased nor impressed.

The last straw was a fire that occurred at the new Kach' aerodrome in July of 1912. As a good deal of damage was done to equipment and buildings that had only recently been completed at great cost, Col. Odintsov was put into a difficult position. Nevertheless, he was as usual undeterred, and addressed the situation in a report submitted to the Grand Duke on July 17.[14] In regard to the extent of the damage, Odintsov explained that the specific instructions he had received from the OVF for re-ordering necessary parts would take time, "and for us *time is more precious than money*." He further noted the nasty impression left by Colonel Semkovskii, commander of the Aeronautics Section of the Engineering Administration, whom Odintsov termed "the personal enemy of the school and myself."[15] Odintsov was chagrined that in a meeting of the Department of the Air Force, "at which besides the directors and officials of the school were also outsiders... Semkovskii loudly announced that the school had done nothing for the War Department." Odintsov explained that with the permission of General Kaul'bars he had gone on to refute the "false words" of Semkovskii, demonstrating that all of the supervisors of the Sevastopol school were once its students; that 22 officers and 4 soldiers who had graduated from 1910-1911 were already in aviation squadrons; that the school had graduated another 9 navy and 8 army officers who had since left aviation; and that in this way the school had provided 50 pilots in half a year. He added that one must into account that of 63 students, 33 were already flying, half of whom had passed the pilot's examination.

Regarding the fire, which he called "a fatal combination of circumstances," Odintsov reported that he had ordered an investigation, had urgently summoned a military investigator, and was still awaiting the investigator's arrival. He added that in the meantime the school was proceeding to begin training pilots on the Nieuports, designating 15 officers who already knew how to fly for the first group; however, as the school only possessed two of these aircraft for instruction, the training would be delayed, and more training craft were needed. Lastly, Odintsov felt compelled to add that "the nerves of several supervisors cannot withstand the intense training and it is necessary to give them 2–3 day rests. My assistant, Capt. Prince Muruzi, is such a nervous wreck that I offered him a two month leave."

In a subsequent report to the Grand Duke three days later, Odintsov was considerably more perturbed as it became increasingly apparent that he would be called to account for the fire.[16] He began by noting that "in the journal of meeting #54 of the OVF, from the very beginning to the end, the dominant theme is the most extreme dissatisfaction with me. There also is regret and condemnation and censure. All of this was strewn with a generous hand, in advance, not having completely positive facts of my guilt." Odintsov defended himself, asserting that "as regards the fire, all measures dependent upon me were taken." Addressing the issue of the Nieuports, he noted that "the Engineering Department, hostile to us, completely controlled the opinions of Baron Kaul'bars, who in that spirit stated his opinion to me that the flights of Dybovskii and Andreadi were scandalous and it was better to remain silent about them." Odintsov, however, defended the flights:

> both pilots received an assignment: to complete an aerial journey, and they executed this marvelously. They confirmed that the Nieuport, selected by Your Imperial Highness and entering the supply of our army (160 of them), is a completely satisfactory airplane and that we, carrying it out, did not make a mistake.

He then turned to equally troubling matters concerning the school's activities, officers on his staff, and the civilian press:

> recently in the newspapers *Novoe vremia* and *Vechernee vremia* there has begun an entire campaign against the school. With great regret it must be revealed that this comes from the innermost depths of the school. Boris Suvorin is a friend of Prince Muruzi and Viktor-Berchenko. Both of them did not sympathize with 1) the transfer [of the school] from the city and 2) the selection of the Nieuports... Such a difference of opinions and the excessive conversations with correspondents by Prince Muruzi force me to utterly renounce the idea of seeing in him my assistant. I must think about a replacement.[17]

Despite all of this, Odintsov reported that training exercises were proceeding normally, and that "as regards the internal order of the school, all is growing stronger and stronger"; he proposed that by the fall all would be completely organized in this aspect, "despite the fact that [the school] has an officer and soldier staff which by its diversity is extremely difficult for administration." As always Odintsov closed by stating that he respectfully had the honor of being the Grand Duke's faithful servant.

In spite of, or perhaps because of, such explanations, Col. Odintsov was subsequently sacked as commander of the Sevastopol School. Worse, he himself was eventually held responsible for the fire and was forced to face judicial proceedings in early September 1912. True to his character, Odintsov did not depart quietly. The proceedings were covered by most of the major newspapers of the time, with a generally favorable attitude to Odintsov to boot, no doubt a source of some embarrassment to the Department of the Air Force.[18] An article in *Rech'* provided a detailed summary of the whole affair.[19] After relating how Odintsov had been accused by the OVF of not taking necessary measures to prevent the fire, and was the only one to be brought to trial, the article noted that at his trial Odintsov had instead laid the blame for "the gaps and deficiencies in the school's organization" at the feet of the OVF itself. He had pointed out that the school entrusted to him was "a disorganized unit" and that to the present time he had received "no instructions, directions, or orders which concerned the supervision of all matters entrusted to him." Further, Odintsov observed that although according to article 35 of the regulations of the OVF he bore full responsibility for the management of all affairs at the school, he was nevertheless completely subordinate to the OVF and without its permission could not at times make necessary expenditures, even from sums already allotted. The article concluded that all of the correspondence regarding Odintsov's projected measures indeed testified to the fact that the OVF, having laid full responsibility on Odintsov, did not vest him with the corresponding power.

A somewhat different account was provided by an earlier article entitled "Who will be the commander of the Sevastopol Aviation School?" in *Vechernee vremia.*[20] Regarding Odintsov's removal, it was observed that

> this measure, associated with the name of one of the best Russian military aeronauts, has given rise to many rumors. We have been informed from a knowledgeable source that the removal of Colonel Odintsov from his post was caused by the great independence of the former commander of the aviation school, who did not always take into consideration the orders of the OVF.

The article continued that it was still unknown whether the school's temporary commander, Capt. Muruzi, would remain at that post, noting that the regulations of the OVF specified that the commander must be a senior officer. In light of this, "the immediate promotion of Capt. Prince Muruzi is doubtful" and according to their source the question of a new commander would be decided upon the return of the Grand Duke Alexander. In light of Odintsov's allegations regarding Muruzi's intrigues, one is forced to wonder whether their knowledgeable source was none other than Muruzi himself. In a later report in *Novoe vremia*, it was noted that the vacant post of commander of the school had been offered to General Staff Colonel V. N. Gatovskii, who had completed the practical course at the School in 1911. He had refused to take the position, however, choosing instead to continue his service in the Officers' Cavalry School. Capt. Muruzi therefore continued to occupy temporarily the post of commander.[21]

Thus ended the aviation career of Col. Odintsov. Ironically, after the removal of this thorn in their sides, the authorities of the War Department and the Department of the Air Force continued to have problems with Odintsov's permanent replacement. This was none other than the newly promoted Lieut. Col. Prince Muruzi. Despite predictions that his quick promotion was unlikely, the support of the Grand Duke not surprisingly made this possible. Once ensconced, Muruzi set to work. Demonstrating that conviction, impatience with inertia, and single-minded devotion to aviation were perhaps contagious, he soon earned the ire of his superiors as well. Emblematic of Muruzi's conduct was his report, marked secret and dated January 21, 1913, to the Department of the Air Force.[22] In it he noted that while the new airplanes of the 8th Aeronautical Company in Sevastopol had arrived as scheduled, and that favorable weather facilitated their delivery to the School's aerodrome, the planes had remained with the company. Muruzi observed that "apparently the commander of the 8th Company does not wish to take any measures to give the officers assigned to the 8th Company the opportunity to fly." He further complained that this commander, knowing that this new aviation squadron would be based at the School, seemingly had plenty of time to discuss matters of organization with Muruzi, "but Colonel Likhachev obviously does not consider this necessary." Muruzi could not hide his frustration with this state of affairs, "especially in view of the anxious times which we are experiencing, when we can expect war and officers will have to fly on aircraft still untested by them in the air, with unfamiliar engines." Muruzi added that "I personally, knowing that in order to fly well as much practice as possible is necessary and that the art of flying is acquired only through practice," had informed Col. Likhachev that in order to give the officers of the 8th company flight time the School would lend him, pending the release of War Ministry credits, gasoline, oil, a storage place for one airplane and quarters for officer-pilots and soldiers; "but I received no reply whatsoever." Muruzi thus concluded that "such an inexplicable attitude towards aviation on the part of Colonel Likhachev cannot help but affect, in a dispiriting way, not only the officer-pilots of the 8th Company, but even the officers of the School, as there is beginning to develop among them the belief that the commanders of the aviation centers, not being aviators, will treat aviation with prejudice."

General Shishkevich's less than demure reaction to this matter was contained in his letter of February 13, 1913, to N. F. Fogel', manager of affairs for the Department of the Air Force.[23] Having been forwarded a copy of the above report, a telegram from the commandant of the Sevastopol fortress further informed him of Muruzi's subsequent proposal to assign all of the airplanes and aviation personnel of the 8th aeronautical company to the Sevastopol School. Shishkevich thus felt compelled to re-iterate in no uncertain terms that direction of aeronautics was "in the Russian army entrusted to me." He emphasized that "the person responsible before the Tsar

Emperor is the War Minister, from whom I receive instructions through the Chief of the General Staff, in accordance with which I should direct my measures." Regarding the dispatch of officer-pilots to the aviation school, Shishkevich proclaimed that "I do not share the fear of Lieut. Col. Muruzi, that pilots graduated from the school in December would forget how to fly by March. I would also recommend this to the aviation school of the Air Force, as pilots trained in the Gatchina School do not forget how to fly during the winter period." Shishkevich concluded by noting that while he always sought to meet all of the wishes of the OVF and executed without delay all measures that depended upon him,

> I cannot allow the interference of Lieut. Col. Prince Muruzi in the sphere of my competence, as such interference, giving rise to temporary, random orders, would disrupt the systematic character of the work planned by me and could again lead to those undesirable relations between the Department of the Air Force and the War Department which existed under Colonel Odintsov.... I ask you not to refuse to inform Lieut. Colonel Prince Muruzi that, appreciating his desire to come to the aid of the officer-pilots of the future Sevastopol aviation company, I nevertheless reserve to myself the supervision of the activity of the personnel of the aviation units and I cannot allow adjustments to this supervision on the part of persons not well-informed of the general course of work in the organization of aeronautics, and I do not have time to give detailed explanations to every interference as is taking place in the present case.[24]

The manager of affairs for the OVF and adjutant to the Grand Duke, Capt. Fogel', was not surprisingly put in a somewhat awkward position by all of this. He was left the delicate task of mollifying Shishkevich, rationalizing Muruzi's behavior, and defending the reputation of the Sevastopol Aviation School. In his reply to Shishkevich two days later,[25] he explained that he had immediately asked Muruzi what could have caused him to send such a telegram to the Commandant of the Sevastopol fortress. For his part, Fogel' proposed that "the sending of the aforementioned telegram was merely the result of private conversations, caused by the rumors circulating and the impending general mobilization." Regarding Muruzi's desire to provide flight time to officer-aviators, Fogel' noted that he could not grant that Prince Muruzi "wished by this to interfere in the command of Your Highness, and if any blunder was made by him, it was only from the desire to be helpful, and by no means to interfere with Your prescriptions" regarding the development of the aviation forces. Fogel' felt it necessary to reiterate that

> regarding officer-pilots who have finished the Sevastopol Aviation School, I am completely sure that if they have to, due to circumstances not depending upon them, remain for a prolonged period of time without practice, that this in no way could have

> an effect upon their ability to fly, as attendance at the Sevastopol school provides an opportunity to rather thoroughly study the matter of aviation, and I am sure that in the future the officers of the last two graduating classes will maintain the glory of their comrades who finished the school earlier and who in the maneuvers of the Warsaw military district in 1911 and 1912 by their service brilliantly proved the benefit brought by them to the army.

Fogel' added that the Grand Duke, "highly valuing Your willingness to always meet all the wishes of the Department, would be very distressed if on account of a few blunders of the Commander of the school, there could develop undesirable relations with the War Department." Fogel' therefore concluded that he would carry out Gen. Shishkevich's orders and direct Prince Muruzi "to in no way enter into any private conversations at all concerning aviation" so as not to harm that cause "to which the August Chairman and all of the members of the Department have devoted so much." He closed by asking Shishkevich to take assurance in his deep respect.

Despite Fogel's posture of deference and good will, a hint of perhaps a very different attitude toward Shishkevich and his competence was provided in a letter of Vasilii Rebikov to the Grand Duke Alexander in June 1912.[26] After his meeting with Chief of the General Staff Zhilinskii, Rebikov had the next day met for the first time with Shishkevich, recently selected to head the new Aeronautics Unit. Rebikov reported that a conversation of approximately 45 minutes concerned exclusively matters of technology, clarifying that "to be more precise it was not a conversation, but the General lectured me on the selection of apparatuses and several times put me in an awkward position; to object was awkward as it could embarrass him but to listen without objection to what he was saying was harmful for the matter, as my silence could convince him in his ideas." He noted that he escaped this predicament by pointing out an example of the potentially unfortunate consequences of the pressure to buy aircraft for the support of domestic industry on which officers would then have to fly. He then remarked to Gen. Shishkevich that

> the matter of the selection of aircraft is very difficult; one must know them marvelously and I think that Your Excellency will not wish to take upon himself the responsibility for the life of officers, who will be required to fly in poor weather. I do not wish to play before you the role of an authority, which title I still do not deserve, and thus I will not insist upon my opinion nor force it upon you, but you have at your disposal an efficient, talented, and rather knowledgeable officer, Capt. Vegener- let the responsibility lie upon him and demand from him an explanatory report...

Rebikov told Gen. Shishkevich that it would be his responsibility whether or not to agree with Capt. Vegener's conclusions. Rebikov further noted that "it is interesting" that the entire time Shishkevich did not question him but instead told him of refinements in Nieuport and Farman aircraft, which Rebikov confessed he knew nothing about, as did the firms which he then contacted by phone; "obviously he [Shishkevich] confused something." In light of all of this Rebikov glumly concluded that "in the presence of the desire of General Shishkevich to supervise the selection of apparatuses, it will be bad for the cause of Russian aviation; all hope is on Vegener, who is known by the commander of the Technical Department."[27] Capt. Vegener of the Officers' Aeronautics School was indeed a knowledgeable and active aviation enthusiast; upon the creation of the Aeronautics Unit he had been appointed as a consulting engineer to this institution. Obviously Rebikov valued knowledge and experience over rank and seniority.

Uneasy relations and a mutual lack of confidence were not exclusive to the Department of the Air Force and the Aeronautics Unit. General Kovan'ko of the Officers' Aeronautics School also had a seemingly uneasy relationship with General Shishkevich. As in the disputes between Odintsov and the Engineering Department, as well as Muruzi and Shishkevich, disputes about policy were often interspersed with subtle and not so subtle appeals to superior expertise and competence. In a small but revealing incident in June 1913, General Kovan'ko was forced to defend himself from an allegation by Shishkevich that one of Kovan'ko's staff was the source of another press leak. The article in question, printed in *Vechernee vremia*, concerned a flight instructor of the Wright-Mons firm who had been assigned to the Officers' Aeronautics School, but who was rumored to actually be an Austrian officer. The basis of Shishkevich's conclusion was that Kovan'ko had used an identical expression regarding this instructor as had appeared in the newspaper article. In a secret reply to Shishkevich, Kovan'ko informed him that the information concerning the instructor had been communicated to him by two of his officers.[28] He noted that "calling the attention of Your Excellency to the fact that in aeronautical terminology the expression 'instructor' applies to all persons teaching piloting on flying apparatuses, it was in this very sense that the expression was used" in one of his previous reports. He added that Shishkevich "in all probability" had attached the same meaning to the expression in one of his previous messages to Kovan'ko. Kovan'ko continued that in his conversations with Shishkevich and in a written missive he had addressed this affair in all possible entirety, explaining that neither he nor his subordinates was the source of the leak, "after which I thought the matter closed." Regarding Shiskevich's allegation, "I ask you to inform me for a report to the Chief of the General Staff which facts You have at your disposal" for making such a determination.[29]

In a terse reply, Gen. Shishkevich noted that the phrase regarding the instructor in the *Vechernee vremia* piece and the repetition of the same phrase in a previous Kovan'ko message "gives me the right to draw a conclusion... to Your Excellency regarding the source of the aforementioned item."[30] He also observed that "the reference of Your Excellency to aeronautical terminology for explanation of the word 'instructor' does not hold persuasiveness for me, as being also well acquainted with such terminology, I do not find that the word instructor has any special significance." Whether Kovan'ko's explanation represented helpful clarification or a subtle dig, Shishkevich was quick to assert his own knowledge and prerogative.

This incident was not the end of General Kovan'ko's troubles with *Vechernee vremia* and its sister paper, *Novoe vremia*, the leading conservative daily in St. Petersburg. As elsewhere in Europe, those on the right were among the most active, strident, and impatient supporters of aviation and its development. As such, Col. Odintsov also had run afoul of the fulminations of these papers and their editor, Boris Suvorin. Unfortunately for General Kovan'ko, they had decided that he too was a fundamental hindrance to progress. Thus, two articles highly critical of Russian military aviation duly appeared in *Vechernee vremia* in June 1913.[31] Primarily setting their sights upon Kovan'ko, they accused him of, among other things, intrigue, arbitrariness, nepotism, and resistance both to the transfer of aviation from the jurisdiction of the Engineering Administration and the opening of the school to officers from non-engineering branches of the army.

As salacious as the charges were, more significant was the series of accusations and recriminations that the articles subsequently provoked amongst the officers of the Officers' Aeronautics School and the Aeronautics Unit of the General Staff. The same contentious issues of professional identity, institutional rivalry, and personal pride that characterized the correspondence between Col. Odintsov, Col. Muruzi, and Gen. Shishkevich surfaced once more. Emblematic was Gen. Kovan'ko's indignant response, which he provided in several secret reports to Generals Shishkevich and Zhilinskii. In a report to Zhilinskii, dated June 19th, Kovan'ko addressed the matter in detail.[32] He noted that the first article, entitled "Airplanes, Pilots, and the Command," was signed with the initials 'V. L.', which supposedly indicated it had been written by a certified military pilot; for Kovan'ko, this rendered the pretensions of the piece that much more disgraceful. Yet he observed that it "contains attacks upon the War Department as regards the allegedly incorrect organization of the matter of military aviation in Russia," which, however, "does not go beyond the bounds of the usual attacks upon the War Department in the spirit of *Vechernee vremia*." The second article, "The 25th Anniversary of our Military Aeronautics," was an entirely different matter, however. Kovan'ko noted that it was an anonymous article which contained, besides selective and incorrect information,

> also a series of attacks upon me personally, upon the teachers of the school and its permanent staff, which supervises lessons with officers, and equally a series of insinuations of arbitrariness in admission to take the course, with the application of malicious will, and an accusation of antagonism, which harms the matter, between the Engineering Administration and the General Staff. It casts aspersions on the unproductive expenditure of money, it contains also a reproach of nepotism.

Kovan'ko thus reported that he was including a copy of the article and for his part requested: 1) orders for bringing a suit against the paper's editor, B. A. Suvorin, through the organs of press surveillance; 2) "permission for me to bring the same editor to trial for insult to me by those expressions, which can harm my honor, dignity, and my good name" through a direct appeal to the procurator; and "permission for me to resort to personally challenging the editor of the newspaper *Vechernee vremia*, B.A. Suvorin, to a duel, if such can consider a duel with a general and a nobleman worthy."

Zhilinskii succinctly replied the next day. Regarding point 1, he declined to bring suit, while regarding points 2 and 3, he noted drily that Kovan'ko had proposed two ways out of the situation "which are mutually exclusive." Thus, choosing not to discuss the measures proposed in point 3, "I permit you to bring the publisher of *Vechernee vremia* to legal responsibility for that part of the article, which touches upon You and the activity of the school entrusted to you."[33] Apparently Zhilinskii did not consider a duel in the streets of St. Petersburg between a noted newspaper editor and the commander of the Officers' Aeronautics School as likely to burnish the public image and standing of the army.

For Gen. Kovan'ko, however, apparently this was not enough. In a subsequent report to Zhilinskii several days later, he provided a detailed rebuttal to the charges levied by *Vechernee vremia*, including numerous direct quotations from the article itself.[34] This report, he noted, was to serve as the basis for a published refutation in *Vechernee vremia* in accord with article 114 of the Statute on Censorship and the Press. First countering charges of huge funds spent on a handful of trainees, Kovan'ko noted that "in reality the staff of the school totals 30 persons and the very modest amount of less than 20,000 rubles is allotted to the school, and not the 'hundreds of thousands of the people's money' as is stated in the article below." The article further noted that the School's staff

> 'this year totaled 28, with 18 persons finishing the course. How is this occurrence to be explained? I will note it with one word- <intrigues>. For example, last year the

> Main Engineering Administration, displeased that the aeronautics unit was being transferred to the General Staff, reasoning as far as they were concerned <after us, the deluge>, not only made no orders regarding new admissions, but together with General Kovan'ko began to reject incoming petitions under various pretexts.'

This, according to Kovan'ko, was "completely fictitious and slanderous," as in accordance with the regulations of the School, enrollment of officers was conducted on the basis of their official documents, namely service lists, attestations, and medical examinations. Kovan'ko asserted that this

> guarantees the impossibility of any violation of law or 'intrigue' of any kind in the enrollment itself; moreover the granting of enrollment in the schools depends on the Aeronautics Unit of GUGSh, and earlier on the GIU; by virtue of this the Commander of the OVS could not and cannot in any way 'reject incoming petitions under various pretexts.' The expression regarding the Main Engineering Administration, 'after us... the deluge,' is a completely groundless and insulting prank against the Engineering Administration.

Kovan'ko continued that "further attacks have the same tendentious tone." Thus, the newspaper further charged that "'while thousands of the most talented and most energetic young officers are trying with all the strength of their hearts to study aeronautics, to get into aviation'" only 28 were enlisted, and of these 10 dismissed. The article's reasoning was that:

> 'they decided in the school to prove the fallacy of the decision which opened access to the school to all branches of arms, they decided to convince that the school was and should remain exclusively engineering. But, for heaven's sake, take exclusively engineers, admit them by exams, make whichever choice you wish, but from personal scores of the command do not destroy hundreds of thousands of the people's money, do not kill time, the loss of which is like death, do not deny Russia.'

Kovan'ko responded that "these attacks completely do not correspond to reality and are tendentious, false allegations." Disputing the existence of "thousands" of potential applicants, he noted that "if one takes into account that besides aptitude and energy, even the outstanding must still satisfy the demands of finishing the course of a military *uchilishche* in the first rank and certain medical standards for service in the aeronautical units, then it is clear that the circle of people" eligible for service was greatly reduced. To the contrary, he added, during the current year the number of those seeking admission was actually less than the norm, attested to by the fact that several officer trainees had been accepted by special permission, even though they had completed an *uchilishche* in the second rank; "this particularly clearly emphasizes the groundlessness of the claims of the author of the article."

Regarding the dismissal of the ten officers, Kovan'ko explained that three had been transferred or dismissed at their own request, while the other seven officers, who had made poor progress throughout the year, received unsatisfactory marks on their exams and were thus dismissed by the Commander of the School, based upon the resolution of the training committee, all of which was done according to school regulations. Kovan'ko further observed that

> it is interesting to note that in regard to the dismissed officers, the command of the school showed special attention, having brought before the Chief of the General Staff a petition about the assignment of four of the dismissed officers, who had displayed love of the matter but who had poor preparation upon entering the school, to aeronautical units, in order to give these officers the opportunity to better prepare themselves and again in the future to enroll in the school, and in the places of their assignment to practically familiarize themselves with the matter of aeronautics.[35]

He emphasized that "there could not and cannot be any kind of arbitrariness or malicious intent," as evaluation of trainees' knowledge was conducted by commissions consisting of 4–5 people, with the participation of a representative of the General Staff; final examination of the grades and progress of officers was conducted by the training committee of the school, which consisted of all teachers and supervisors of practical lessons.

Kovan'ko asserted in the same vein that "the statement of the newspaper about the desire of the school to preserve it as exclusively engineering is groundless and does not correspond to reality," adding that "no personal scores between the command of the school and the administration were observed, indeed could not be as a result of the clear, sharp delimitations in the sphere of activity of each." Assignment to the school did not depend on the school itself, "but instead was a function of the Administration/the Aeronautics Unit or in the past the GIU," while the supervision of the training course and the exercises of the officers "constitutes the concern of the school only." He emphasized that "the common efforts of the commanding persons are directed towards the best execution of a common cause—the training of good aeronautical officer-technicians."

Regarding the series of attacks against the staff and personnel of the school, Kovan'ko observed that "these attacks are slanderous and they can most likely be attributed to the category of gossip," noting that they were of very minor significance compared to other allegations raised in the article. However, one of these, no doubt particularly awkward for Kovan'ko, concerned the charges of nepotism in regard to his son, Lieut. Kovan'ko, a member of the permanent staff of the OVS. Kovan'ko noted that the article's author cited this "'piquant fact'" which portrayed the arbitrariness that allegedly reigned in the school. Kovan'ko vigorously retorted that "this fact, or to be more precise a malicious attempt to accuse the command of the school of nepotism, is from beginning to end a lie." He demonstrated in detail that all vacancies in the staff from 1910 had been by the next officers in line; when the school's Aviation Section had been formed at the end of 1912, four vacancies had opened, which were again filled by the next officers in line, "of which one was Lieut. Kovan'ko, the son of the commander of the school, who in 1911 had finished the school second in the list out of '28' officers and who had been assigned to the school for 1½ years, of which more than ½ a year as instructor in the Aviation Section."

In conclusion, in light of all the evidence adduced, Kovan'ko argued that there were sufficient grounds both for the initiation of an official prosecution of libel and for the drafting of a refutation. Moreover, he informed Zhilinskii that "in general the article by its tendentiousness and by its knowingly false information outraged the officers of the school, who appealed to me with a request to allow them a general assembly for the discussion of the article." He noted that by his order the general assembly had been permitted and that he was submitting along with his report an excerpt from the assembly's resolutions. Kovan'ko then made several formal requests of Zhilinskii. He first asked for permission to compose a refutation based on material in the report, which together with the resolution of the general assembly of officers was to be published in *Vechernee vremia*, all on the basis of the Statute on Censorship. Further, in accordance with guiding instructions of the Chairman of the Council of Ministers, he asked to strengthen the text of this refutation with the statement that the editor of *Vechernee vremia* was being brought to legal responsibility "a. for libel and insult of the dignity of the military sphere [*voinskoi chasti*]," and "b. for libel, which is injurious to the honor, dignity, and good name of the Commander of the OVS, General-Lieut. Kovan'ko." Lastly, Kovan'ko asked for permission to publish the refutation and the excerpt from the resolutions of the assembly of officers in the newspapers *Russkii invalid* and *Russia.*[36] Apparently Kovan'ko did not want the indignation of himself and his officers, nor an explanation of the true state of affairs at the OVS, limited to the readers of *Vechernee vremia*.

Personality and Professionalism

What to make of all this? After the examination in preceding chapters of grand visions and comprehensive projects to develop aviation, this extended foray into the minutiae of institutional rivalry, personal bickering, and petty intrigue is nevertheless instructive. As was the case with the plans to develop aviation, the numerous and often heated differences of opinion were responses to very real and fundamental problems posed by a new and complex technology. As pointedly noted by both officers and pundits in the press, upon such policy decisions as the selection of aircraft types the lives of actual pilots depended. This weight of responsibility was magnified by the sense of urgency generated by the rapidity of international aviation development and the enormous possibilities envisaged for the airplane. It was against this backdrop that the commanding officers of the aviation services sought to assert leadership and exercise daily management over the course and implementation of aviation policies. Yet their competing efforts to do so often flared into dissent, accusation, and even confrontation; their communications were marked by charges of obstruction, interference, and presumptuousness. In their relations these officers implicitly and explicitly invoked claims of expertise, prerogatives of rank, and perquisites of honor; they cited both their functional competence and moral standing, even if it meant disparaging such claims by others. As such, they appealed to two competing paradigms to assert and justify their authority: professionalism and personality.

As discussed earlier, in the aftermath of defeat and the often violent repudiation of established structures of hierarchy that accompanied the Russo-Japanese War and Revolution of 1905, numerous elements of the senior army leadership were believed to have forfeited their privileged status and prerogatives. Both prominent individuals and ostensibly elite groups, including the General Staff, had all been shown to be wanting in their performance. For such groups subsequent efforts to reassert their authority were complicated by lingering weakness, both moral and material. Moreover, they were forced to contend with competing attempts by other elements of the officer community to modify or reconstitute patterns of legitimacy and authority. These issues occupied a crucial place in the discourse regarding the limits and meanings of reform within the army. Numerous figures across the spectrum of these debates increasingly appealed to more explicitly professional models of service, framed around rigorous education, thorough preparation and training, functional specialization, and the sense of identity and camaraderie derived therein. Alongside the cultivation of a professional ethic, however, numerous officers within the army continued to assert that an officer's moral authority to lead was not reducible to objective claims of expertise and abstract norms of competence. Instead, they argued for the importance of what might loosely be termed the role of "personality." Individual genius, personal integrity, devotion to the sovereign, commitment to uniquely Russian traditions, or social status could all be construed as integral aspects of an officer's identity. Such aspects, which did not necessarily inhere in his specific role and place within the military hierarchy, became the locus of an officer's authority and prerogative. Hence in part the emphasis on personal honor, and the exquisite sensibility to insult real and perceived, that was such a noted and lamented feature of the Russian officer corps in this period.

As is readily apparent, although distinct in the abstract, and to an extent antithetical as a means of legitimizing rank and authority, in practice elements of professionalism and personality were often difficult to disentangle. William Fuller notes too that professional and corporatist conceptions of role and duty could indeed coexist in any one officer.[37] However, I would argue that the inherent tensions between professional and what might be termed "patrimonial" paradigms of service were manifested most strikingly in the absence of a neutral sphere to manage issues of disagreement, dissent, and criticism. In contrast say to the traditions of the German General Staff, which encouraged collegial and often brutal criticism amongst commanders and subordinates, in the Russian army differing policy positions, conflicting appraisals and opinions, and the routines of peer review frequently if not inevitably became personal issues, and thus issues of personal honor. This strongly colored relationships ostensibly framed around professional functions. Likewise, in the absence of fully normalized procedures or channels to debate issues or express dissent, intrigue became politics by other means. Politics is not always devoid of policy, just as intrigue is not completely devoid of content and real policy differences.[38]

These manifestations of the complex and often tense relationship between professional and patrimonial paradigms of duty again took on a unique cast within the aviation service. Distinct sub-cultures of professionalism, variously rooted in the General Staff, the engineering department, and the aeronautical service, found their expression in the commanding officers of the aviation schools and institutions. While claiming to share a common vision of a powerful air force, these officers appealed to their own standards of competence, expertise, and authority as a means to realize that vision. This was in part the source of friction among them. Yet in a branch of arms largely bereft of a lengthy history, established tradition, or institutional culture of its own, the importance of individuals, particularly those in positions of responsibility, loomed large as well. Competing professional paradigms were thus further distilled through the prism of personality. The clashes, intrigues, factional rivalry, and slander real and imagined that frequently marked relations amongst these men were skirmishes in a struggle over the future identity of the nascent air force. The means and channels through which this struggle was fought were varied. The prestige of a Grand Duke, the imprimatur of the General Staff, a respected aeronautical tradition, were all invoked as compelling sources of competence and prerogative; the use of patronage, the unilateral exercise of initiative, the manipulation of the press, resorts to personal attacks, were all considered fair game.

In regards to the tension between professionalism and personality, the problematic relationship between authority, competence, seniority, and rank was perhaps the most contentious issue in the realm of military aviation. For some leading officers, their commitment to their cause, their sense of urgency, and their recognition of the enormous stakes involved all combined to create a situation where established patterns of routine, custom, and hierarchy were seen to be particularly cumbrous or even galling. The rapidity of the airplane's development and the scarcity of those with requisite expertise put a premium upon flexibility and responsiveness, attributes for which the tsarist army was not noted. Leading officers thus often chafed at being at the mercy of superiors and bureaucrats with no experience and little practical knowledge of aviation. Due to the rapid development of the airplane, there was a dearth of senior, experienced personnel to fill positions of leadership. In fact, there were only a handful of persons above the rank of captain with both firsthand knowledge of aviation and practical training as pilots. This was an important distinction, as there were numerous knowledgeable enthusiasts who held high rank and served in the aviation services, but few of them had actually flown.

Fundamental as it was, this indeed was not a problem exclusive to the Russian army; all of the European air forces grappled with the challenge of reconciling expertise, experience, and rank in a newly emerging branch of arms.[39] However, the unique institutional structure of the Russian air services exacerbated it and complicated the patterns and channels through authority and leadership were expressed. Thus, as noted, Gen. Shishkevich was charged with the entire direction of military aviation when in fact he was not a pilot. Yet various officers like Colonels Odintsov and Muruzi as well as pundits in the press questioned whether non-aviators could adequately represent the occupational interests and concerns of a cadre of officer-pilots. Gen. Shishkevich's frequent defensiveness and zest to assert his prerogatives in his dealings with others suggest that he certainly believed so. Always jealous of his authority, he was not shy about asserting the preeminence granted him by his position, nor was he above appealing to institutional rivalry. The brief and tempestuous aviation career of Col. Odintsov further illustrated how the very youth of the aviation service exacerbated the influence of personality. While he had received some training as a pilot, Odintsov apparently could not fly independently as a fully certified aviator; the same was true of Muruzi. Yet it was Odintsov's fervent and single-minded commitment to the expressed goal of the Grand Duke, that of building a powerful aerial fleet as rapidly as possible, that seems to have blinded him to the nuances of routine, office politics, and interpersonal skills, ultimately leading to his removal.

At the same time, the disparate experiences of Shishkevich and Odintsov point to the fluid meanings of professionalism as an ethos of service was framed around the airplane. Both were officers of the General Staff, and as such ostensible beneficiaries of the same rigorous education, professional ethic, and strong esprit de corps. However, each chose to express his devotion to aviation in very different ways. In contrast, the continuing lure of the cavalry as an elite branch apparently caused General Staff Col. Gatovskii to refuse the post of commander of the Sevastopol School when given the chance. The respective career paths of these officers illustrate that under the rubric of the General Staff was a diverse collection of individuals, united by a common process of examination, admission, and training, but nevertheless with a variety of personal desires, beliefs, and conceptions of where their duty lay. The latter was true of the officer cadre of the aviation service as a whole.

The tangled relationship of personality and professionalism, authority and competence, was perhaps most evident in the commander of the one aviation institution that did have an established history: the Officers Aeronautics School. Gen. Kovan'ko had been instrumental in the school's founding over two decades before, had served in command throughout its entire existence, and had over the years appointed his proteges to subordinate positions. It was to this tradition, and the authority it conferred, that Kovan'ko often appealed in his letters to Gen. Shishkevich. Yet while Kovan'ko had to an extent welcomed the transfer of aeronautics to the jurisdiction of the General Staff,[40] his vigorous defense of the efforts of the Engineering Administration and particularly his own School indicate that he saw the Aeronautics Unit as an administrative partner in the cause, but no more. The strained tone of the resulting correspondence between the two men demonstrates that their relationship was not an amicable one. At the same time, it was the pervasive imprint of Kovan'ko's personality on the Officers' Aeronautics School that often caused his critics to rail at him as a fundamental hindrance to further progress, as he too was not an aviator, but an aeronaut.

The clashes between senior aviation officers and their various critics vividly illustrate another key feature of the uneasy relationship between military professionalism and personal authority. This was the continuing and powerful role of honor in defining the behavior and self-perception of officers. If the inability to separate professional opinions from personal insult was a problem within the confines of the officer community, it was further complicated within the context of a newly liberated and increasingly assertive civilian press.[41] As emphasized in previous chapters, the proposed partnership between tsarist military and Imperial society in the development of the airplane was always an ambivalent one; pleas and hopes for public support were accompanied by fears of civilian interference and criticism. Such ambivalence was clearly manifested in Gen. Kovan'ko's outrage at the editors of *Vechernee vremia.* Aspersions upon his professional competence and behavior were transformed into questions of personal dignity, culminating in a request, as "a general and a nobleman," to shoot the editor Suvorin. A similar if less intense tone of personal defensiveness prevailed in the dispute between Kovan'ko and Shishkevich regarding alleged press leaks, as well as in Col. Odintsov's acrimonious relations with the Engineering Administration. For an officer corps allegedly ready to kill over the slightest infringements of honor, such attitudes were not uncommon, and the disagreements in question were relatively tame. However, they reinforce the point that even for those tsarist officers who were well-educated, highly trained, and thoroughly professional, there were few avenues for communication and especially criticism which did not ultimately involve matters of personal honor.

Lastly, the complicated dynamic of personality and professionalism was manifested in two men who were rarely directly heard from, yet who lurked like shadows behind the disputes examined in this chapter: Minister of War Sukhomlinov and the Grand Duke Alexander Mikhailovich. They were instrumental figures in creating the institutional and cultural framework that was to guide the development of Russian aviation. Sukhomlinov, once seen as little more than a corrupt schemer and indifferent policy maker, has recently been subjected to a reappraisal. If he was not a visionary, he was not quite the reactionary once portrayed in the literature either. At the least, he was committed to strengthening and modernizing the army, at least as he conceived of it. Both Fuller and Menning have argued that in fact he did a good deal to improve the army.[42] His fiat designating the development of aviation as a priority and his command that this be undertaken by the General Staff bears this out. Although there were longer-term tends at work, and though his decisions were frequently viewed by his colleagues as frivolous and spurious, Sukhomlinov's marked support certainly helped to spur the growth and improve the status of Russian military aviation. This is a rather different appraisal than that provided in Soviet historiography, which notes Sukhomlinov "had no understanding at all of questions of aviation," yet meddled by arbitrarily giving specific, absurd, and ultimately harmful orders in the matter.[43] If anything what Sukhomlinov's role points to—and it seems this was a phenomenon especially marked in the Russian army—is the difficulty in neatly categorizing reformers and conservatives, professionals and dilettantes. Venality and a sense of duty, informed appraisals and personal hostility, ostensibly military considerations and non-military motivations could all happily coexist in a man like Sukhomlinov.

In another sense, one of the more visible manifestations of the uneasy relationship between personality and professionalism was the continuing part played in military affairs by the various Imperial Grand Dukes, who often occupied senior positions in the various branches of the military, and whose influence was obviously not solely reducible to professional considerations or their place in the chain of command. It was this anomaly that motivated a prominent civilian advocate of army reform and professed aviation enthusiast, Chairman of the Duma A. I. Guchkov, to make his extraordinary speech in June 1908, in which he appealed to the Grand Dukes who served as inspectors in the various branches of the army to resign "on the patriotic ground that as members of the Imperial family they were not subject to criticism, and that they thus stood in the way of all proposals of reform."[44] This situation found opponents in the army as well. Particularly vexing to many officers, especially those committed to a more merit-oriented system, was the penchant of the Grand Dukes for surrounding themselves with hand-picked generals, who had often done nothing noteworthy, in their personal suites.

Yet there was perhaps no better manifestation of the vagaries of the influence of Grand Dukes than Alexander Mikhailovich and his creation, the Department of the Air Force. He was a powerful, influential, and arguably effective advocate for the development of Russian military aviation. Not only did he provide moral and material capital, he seemingly came to the matter free of the baggage- institutional interests, a clinging fondness for balloon technology, red tape, apathy- that seemed to hover like a cloud over the War Ministry and the official aeronautics organs. I have already alluded to the discomfort this caused contemporaries and a later generation of Soviet historians. While they could question his pretensions and point to notable blunders, such as the selection of the Nieuports, this appears to be one instance where the activity of a Grand Duke was to an extent positive, progressive, and by its "issue an order and let it be done" nature, effective. Moreover, historians of the Imperial army usually judge the meddling of the Grand Dukes as particularly pernicious because such activities, whether harmful or beneficial, largely existed outside regular army structures and channels, thus becoming matters of personality, not institutions and policy. However, in this case Alexander's efforts were indeed institutionalized in the form of the Department of the Air Force and the Sevastopol Aviation School. The Grand Duke also played a large role in installing his own selections to positions of command; the respective personnel—the leadership of the OVF and the permanent staff of the school—were to largely remain in place throughout the pre-war period. Their unique role as active officers, in the service of a royal patron and quasi-public organization, yet dedicated to the creation of a military air arm, attested to the complex relationship between duty, personality, and professionalism for tsarist officers.

While frequently unseemly and even petty, the politics of command in the aviation services revealed the broader and potentially subversive potential of the airplane to challenge existing paradigms of authority and leadership in the tsarist officer community. The commanders of the aviation services perhaps felt most acutely the responsibility, promise, and requisite challenges that the airplane generated. However, the officers of the line in the aviation services were not merely passive objects in this process. Instead, the aviation schools and units were another fruitful area of activity and frequently contention in thrashing out an ethic and culture of aviation service.

6| Pioneers, Poseurs, Professionals

Officers throughout the aeronautical services, the General Staff, and the army community as a whole grappled with the challenge of assimilating aviation as a new branch of arms on multiple levels. In the initial reception of the airplane, the specific plans to develop aviation, and the debates which swirled around aviation policy at the level of daily command, these officers addressed issues of both technology and personnel, of military priorities and institutional politics, of the realm of the army and that of the state and civilian public. These projects and debates spoke to more than specific matters of organization, administration, and procurement. At times implicitly, at times explicitly, commanding officers in particular sought to prescribe and cultivate a larger ethos of service for the Imperial air forces.

Realizing these visions, however, ultimately depended upon marshaling the human resources who would make the potential power of the airplane a reality. Aviation was a phenomenon of the field and the air, not the marble halls of the ministries and the staffs of the chancelleries. Building the Russian air forces thus revolved around the fundamental challenge of creating almost from scratch a new profession of arms, which soon received the official designation of *voennyi letchik* [military pilot]. This process involved several distinct but closely related tasks: recruiting a cadre of suitable officers, articulating and developing a requisite body of knowledge and skill that would qualify one as a military pilot, and providing a structure of service for those who successfully mastered these demands and joined the community of aviators.

It was a process that proved to be uneven, complex, and often contentious. With few ready analogues within the army to provide a guiding model, knowledge and experience in the recruitment and training of pilots were often by necessity acquired only through trial and frequently lethal error. Equally problematic were the broader questions and concerns that arose as the airplane was rapidly transformed from a novelty into a weapon with the potential for widespread application. What exactly would military aviation entail in terms of demands and commitment? Which groups within the army were thus best suited to utilize and develop the power of the airplane? What would be the nature, purpose, and goals of aviation service? Such questions were confronted by all of the European armies in the development of military aviation.[1] Again, however, they took a unique form within the particular institutional and cultural context of the Imperial army. The competing visions and philosophies of the leaderships of the Officers' Aeronautics School, the Department of the Air Force, and the Aeronautics Unit of the General Staff made themselves felt in the elaboration and implementation of policies of recruitment, training, and service. Such divisions were further mediated within the setting of the aviation schools and units. Particularly for the instructor staffs of the schools, on the sharp end of a new, unperfected, and quite dangerous technology, aviation entailed more than abstract matters of funding, policy making, or institutional rivalries; it involved real people, hard choices, and often matters of life and death. Although often granted a notable voice in the formation of policy due to their invaluable expertise, such officers were not ultimately responsible for structuring the conditions of aviation service. Yet due to their position, function, and status, they had a unique opportunity to articulate and impose their own ideals of such service, ideals which would not always mesh perfectly with the policies and priorities of the army leadership or even the commanders of the schools themselves.

One last constituency would have a say in the creation of a cadre of aviators: the future military pilots themselves. Policies of recruitment, training, and service were ultimately negotiated within the broader social and cultural context of the officer community as a whole. While ostensibly the subjects of regulations devolved from above, officers throughout the army would appraise the opportunity presented by the airplane according to their own particular conceptions of duty and service. Those who seized this opportunity would thus do so with varying motivations and expectations, some of which would bear little relation to the assumptions and goals which underlay official policies.

The complex interplay of disparate visions, institutions, and constituencies would determine how the title of *voennyi letchik* was filled with both practical content and larger meaning, in terms of the specific demands of piloting an airplane and a broader ethos of service that would govern the new officer-pilots. The contentious efforts to articulate the terms and spirit of this novel profession would in turn underpin a still more elusive process, the cultivation of a larger culture of Russian military aviation.

Recruitment and Personnel

Through 1911 the cadre of Russian military aviators was quite small, numbering no more than a few dozen trained pilots. A good number of these pilots comprised the permanent staffs of the Aviation Section of the Officers' Aeronautics School and the Sevastopol Aviation School of the Department of the Air Force. The officers of the former had been culled largely from the aeronautical units and particularly the staff of the OVS, having been trained either abroad or by civilian aviators; those of the latter had been drawn from all branches of arms and had been trained either in France or in the first classes in Sevastopol. While the system of recruitment was neither uniform nor completely regularized for the two schools, by 1911 their staffs were in place and a handful of aviation squadrons were staffed and in existence.

However, the year 1911 brought both the recognition that aviation was poised to grow dramatically as well as concrete proposals to manage this growth. As evidenced in these plans, one of the pressing areas of concern was the problem of personnel, both in terms of quantity and quality. Before a sizable cadre of pilots could be trained and posted, suitable candidates had to be identified and selected. The initial decisions determining eligibility to pursue a career as a *voennyi letchik* markedly influenced the entire subsequent development of the tsarist air service. The most fateful choice in this regard was that of the War Ministry to open the new Aviation Section of the Officers' Aeronautics School to volunteers from all the branch services of the army. Mirroring the policy adopted from its beginning by the Sevastopol Aviation School, this marked a notable shift from the previous exclusively engineering composition of the OVS. A circular report, dated Oct. 26, 1910, of the Aeronautics Section of the Main Engineering Administration to the staffs of the military districts thus spelled out in detail the criteria, procedures, and goals regarding the application process.[2] Specifying the desire to accept officers and soldiers from all branches of arms for pilot training, the circular explained that upon completion of training such officers were to be posted to aviation squadrons of the aeronautical companies, where they would be obliged to serve two years, while lower ranks were to serve in aeronautical units. As per order of the War Minister, the report thus asked the district staffs to submit to the Aeronautics Section lists of officers and soldiers wishing to volunteer on these grounds, along with the conclusions of commanding officers regarding their subordinates' petitions.

The report noted that officers recognized as fit for aviation service were those with sharp vision, good hearing, fully healthy lungs, and sufficiently strong build; a mandatory medical certification of each officer's health was to be sent with each petition. The report added that highly desirable for volunteers was some familiarity with bicycle, automobile, and other types of sport, as well as some familiarity with internal combustion engines used on airplanes. Lastly, such volunteers had to possess good service attestations. For lower ranks the report stipulated basically the same medical criteria. Necessary as well was sound knowledge of metalworking or joiners crafts, while familiarity with combustion or at least steam engines was desirable. The criteria for volunteers were later codified and made even more stringent as the application process was standardized and applied both to the Aviation Section of the OVS and the Sevastopol Aviation School. To be considered for admission to one of the schools officers were required to petition either directly or through the chain of command to the respective school or to the Aeronautics Section of the Engineering Administration, and later the Aeronautics Unit of the General Staff. The latter institutions retained ultimate control over admissions. Officers were to have a rank no higher than captain, were required to have graduated in the first rank from a military *uchilishche*, had to have their most recent service attestation graded "outstanding," and were required to meet strict medical criteria, as well as a maximum age limit of 45 years old for duty in airplanes.[3]

In the order which formally established the Aviation Section, the War Ministry failed to give any explanation for its policy regarding volunteers or evidence of its reasoning. However, the policy clearly reflected the expansion of aviation from a largely technical and experimental matter to one with potential for serious practical application. The War Ministry thus decided that it was useless to continue to pigeonhole aviation as the exclusive domain of engineers, especially as the Sevastopol Aviation School pursued a more inclusive policy. In light of the anticipated rapid expansion of the aviation service, it was sensible to expand the pool of potential applicants. This policy also paralleled the one instituted in France, which again served as the primarily foreign model for officers of the Russian aeronautical services.[4]

The application criteria themselves required nothing more than intelligence, initiative, and ambition, as testified by good health, an exemplary record, and a presumed interest in aviation. In contrast to the engineering officers of the OVS and the soldier volunteers, some of whom were to be trained as pilots, no technical knowledge was required for officer volunteers. In turn, the standards for determining what constituted an exemplary record were high educational achievement (in any branch of service) and an outstanding attestation, which according to tsarist army practice would include a superior's evaluation of both the officer's performance of duty and his moral qualities. These were perhaps the only standards which could be commonly agreed upon to determine the relative merit of officers who would possess disparate education, training, and career backgrounds. These were also perhaps the only standards of merit which could find consensus in an officer community so divided by contending conceptions of duty and service.

In what they did and did not demand, the application criteria thus spoke to an ideal type for volunteer officers and the unique character of their future service. In recognition of this unique character and in order to attract such suitable volunteers, the War Ministry further adopted policies that were becoming standard practice in the air services of Western European militaries: the institution of flight pay, bonuses, and other perquisites of service to attract and retain qualified volunteers; a reduced term of service for a pension; and the provision of pension benefits to surviving spouses and children in light of the dangerous nature of the assignment. The official announcement in *Razvedchik* regarding these measures openly appealed to the special nature of the service of pilots and aeronauts:

> the establishment of the advantages enumerated has a dual significance: on the one hand, it is just compensation for pilots for their difficult work, involving the risk of life; on the other hand, the selfless work of the aeronauts in their dangerous affair is only possible when he sees it is widely valued by the state, when he, knowing what risks he exposes himself to, will be sure that in the event of his death, his family will not be left in a situation worse than that in which they found themselves during his life.[5]

The audience for such appeals was seemingly a large and receptive one. The War Ministry perhaps recognized that it had one more recruiting incentive working in its favor: the stultifying and pointless routine of much of army life. As described in chapter two, the airplane was seen as providing a literal escape from the confines of regimental chancelleries and paperwork; it appeared to offer instead the opportunity for glory and excitement, a sphere for the exercise of initiative and interest, or simply the chance to render meaningful and fruitful service. In the eyes of many young junior line officers frustrated by their lack of authority, the inertia of superiors, and slow rates of promotion, a career in the air seemed to hold a promise and potential that could not be found elsewhere in the army.

Yet despite rigorous standards, patriotic appeals, material incentives, and what seemed a large and willing pool of volunteers, the actual process of recruiting a cadre of officer aviators quickly ran into difficulties. Some of these were a result of the sheer rapidity of aviation development. There were constant fluctuations in the number of officer pilots that the Aeronautics Unit and the War Ministry predicted the army would need, as well as a fluctuating correspondence between the numbers of those trained and training and the actual vacancies open to them in the aviation squadrons. This complicated efforts to both attract and efficiently utilize volunteers.[6] Fear played a role as well in deterring volunteers. As alluded to in the *Razvedchik* announcement, aviation was indeed, despite its attractions, a very "dangerous affair." Clues as to how dangerous could readily be found in War Minister Sukhomlinov's order to select only volunteers who were bachelors. Despite the noble words cited above about ensuring that a pilot's family would be provided for, this proposal was made in order to lessen pension expenditures in the fairly likely event of the death of such officers during service. Confirmation of the hazards of aerial service could be found in an article entitled "Victims of Aviation" which appeared in *Voennyi sbornik* in January 1911; it listed all 28 deaths of aviators around the world in just two short years.[7] Many officers thus apparently concluded that discretion might indeed be the better part of valor.

Difficulties attendant to the recruiting process were far more a product of professional tensions regarding aviation. Although the application regulations ostensibly determined what types of officers were fit to serve as pilots, this official ideal continued to be contested in various forums. Potential volunteers framed their own disparate responses to official policies, as did outsiders, such as the superior officers required to furnish recommendations, who were granted a role in the application process. Further, the application regulations themselves were greeted with lingering dissent, not least among established constituencies in the sphere of aeronautics, but among concerned observers as well. Thus, the decision to accept volunteers from all branches of arms seems to have been motivated as much by sheer necessity as foresight. In response to a July 1907 appeal from the Main Engineering Administration to various engineering units for volunteers for what was then training in aeronautics, the chiefs of engineers of seven fortresses all replied that they had no officers wishing to volunteer. The appeal itself was motivated in part by a 30% shortage—25 officers—in the regular complement of aeronautical units, which was projected to only get worse in the next year.[8] The situation did not change much with the arrival of the airplane. In response to a May 1912 circular from the Engineering Administration regarding volunteers for aviation service, the 5th army corps and the staffs of the Caucasus, Petersburg, and Omsk military districts all reported that they had no volunteers from the engineering units under their command. The 5th army corps, on the other hand, listed ten volunteers from infantry regiments.[9]

This lack of volunteers reflected professional considerations, at least in a negative sense. Many engineering officers apparently considered aviation not only dangerous, but not germane to their training, function, and career path. Perhaps even before the army leadership did so, they recognized the professional distinction between the two services. Alternatively, the May 1912 response may have been influenced at least in part by the impending transfer of aviation from the auspices of the Engineering Administration to the General Staff. As seen in Gen. Kovan'ko's troubles with the press, there was a good deal of gossip regarding alleged resistance to this move on the part of the staff of the Officers' Aeronautics School. The persistent reports of the unhappiness of such officers perhaps bore witness to similar discontent among the corps of engineering officers as a whole. In contrast, there were groups within the officer community who welcomed the creation of the Aeronautics Unit and the new professional direction it would bring to military aviation. This was exemplified most vividly by those graduating officers of the Sevastopol Aviation School who voted with their feet in 1911 and refused to take postings in the aviation squadrons as long as aviation remained under the jurisdiction of the Engineering Administration.[10] This was another instance of line officers putting forth their own ideal of aviation service in the face of official policy, in this case to the vexation of the War Ministry.

This institutional shuffling, however, only served to underscore the uncertain status of aviation as a profession. For the individual officer, reluctance to commit to this new endeavor thus persisted. In his later project for the reorganization of the aeronautical and aviation services, Gen. Shishkevich of the Aeronautics Unit therefore focused upon the fundamental and persistent problem of the lack of career opportunities, rather than fear, as the reason for difficulties of recruiting and retaining officer pilots. He emphasized the problems presented by the lack of positions of senior command and the resulting dead end for promotion this offered most officers, engineering or otherwise. As noted, his response was to recommend the formation of a separate aeronautical corps, which would provide institutional structure and opportunities for a new profession of military aviation.

Such opportunities meant nothing, however, to those who could not take advantage of them. If the culture of routine duty within the army frequently provided the impetus to volunteer for aviation service, that same culture often functioned as an impediment to actually realizing this opportunity. The debilitating habits that indeed plagued the officer community—sheer inertia, short-sightedness, attachment to routine, narrow self-interest—often stood in stark contrast to the visionary ideals for aviation service.[11] Thus, following the Oct. 26, 1910 circular regarding volunteers for aviation service, a subsequent circular of Dec. 11 from the Aeronautics Section to the military district staffs complained that a few districts had sent either only scattered petitions, or the requisite lists of volunteers, but not all of the information specified in the Oct. 26 circular. The report noted that this was complicating the process of posting trainees to the schools.[12] Another report of the Aeronautics Section to the Main Staff on Dec. 28 went further and complained that not one military district had yet submitted a list of officers nominated.[13]

The clash between exalted principles and petty concerns was illustrated by the bachelor issue in particular. A Dec. 1911 report from the Main Engineering Administration to the War Minister therefore pleaded with him to reconsider his order that only bachelor officers be appointed to the aviation schools.[14] It noted that under established procedures volunteers were selected regardless of marital status; thus, 30% of engineering officers who had expressed a desire to volunteer were married, while 10 of 20 already trained pilots were married as well. The report added that in 1912 about 100 officers would be training, and that with the further development of aviation their numbers would only grow; however, with the exclusive use of bachelors,

> the number of officers, from which is now conducted the selection of the best and most aspiring to aviation service, will be significantly reduced. This will lead to not those most able and ready to completely devote themselves to aviation getting into the aviation squadrons, but those bachelor officers to whom access to the school will turn out to be possible only because they have advantages over married officers.

The report observed that if limiting expenditures to the families of dead and wounded pilots was the rationale for the decision, such expenditures would not be large. Thus, "in order not to deprive married officers who have a calling for aviation service" of the right to be trained and serve, the report concluded by asking if it were possible to accept "not exclusively, but primarily bachelor officers." This request was duly approved by War Minister Sukhomlinov.

Incidents like the bachelor controversy further raised the ire of self-appointed critics within the military community who weighed in with their own appraisals concerning the process of recruiting a cadre of officer aviators. Thus, N. Dukhanin, who provided the elaborate critique of Russian military aviation examined earlier, also complained in his article that the staffing regulations of the Officers' Aeronautics School did not provide for "a careful and strict selection." He emphasized that "it is extremely essential that for training in this serious and crucial special service officers be selected with the physical, mental, and moral qualities necessary for it, and with known aptitudes and the appropriate intellectual preparation." He thus proposed establishing theoretical and practical examinations which would provide "a guarantee that the candidate is fit for the study of this matter."[15] In short, he flatly rejected the model whereby generally exemplary service and a curiosity regarding aviation might suffice for acceptance to pilot training.

An anonymous contributor to *Razvedchik* in May 1913 reiterated this theme as he identified numerous persistent shortcomings in the staffing system.[16] He observed that in regard to the development of military aviation, "it goes without saying, that what is necessary for this are workers in the full sense of this word, that is people who have completely devoted themselves to this arm of service, and not those who are only interested in this new field of military science as a fashionable sport, a convenient topic of polite conversations, etc.." However, he noted that while the right to volunteer was ostensibly offered to all junior officers from all branches of service, "this right, with few exceptions" was much of a dead letter, although there were many who satisfied the necessary conditions, both in terms of health "and in regard to an interest and desire to completely devote oneself to the matter of aeronautics." Regarding the dual track petition process, which itself evoked his criticism as evidence of indecision, he observed that successfully petitioning through the chain of command was extremely difficult, "as each unit commander, observing most often only his own personal interests, despite the desire of many of his subordinates to study aeronautics will try in every way either to dissuade them, or, finding the least bit valid excuse, will refuse his agreement." Even if the immediate commander did agree, with the next commander in line there would be the same principle of personal interest, "only in a broader sense." Petitioning directly to one of the aeronautics schools was even less feasible because a unit commander consenting to such a petition would thereby be circumventing his own immediate superior, which "certainly" would not work to his benefit. In regard to units with a significant shortage of officers, a successful petition was not even worth talking about.

Shifting his attacks from the culture of apathy and narrow self-interest that reigned in the army, the author also noted "a whole series of secondary difficulties": the required medical certification with such improbably strict demands "that only a person from the stone age could satisfy them"; the requirement whereby no officers with attestations lower than "good" could apply; and finally, that officers of the Roman-Catholic faith were greatly impeded from learning how to fly. The author recommended instead a system whereby students would be admitted to the schools conditionally and instructors would be allowed to determine whether they satisfied the requisite demands: "an instructor will more quickly and more reliably determine in his student courage, self-control, and caution than a doctor-psychiatrist." Only after an instructor found the "necessary material" in his student would the latter be accepted, while others would be sent back to their units. If a drawback to such a system was increased expenditures, "this is completely justified by the importance of the affair."

In contrast, the author noted that all of the existing problems led to "a very sad result, as many young officers who are actually fit for service in the aeronautical units remain on the flanks, while in the Officers' Aeronautics School to the present time there is a significant shortage" in the complement of trainees. It was not merely a matter of numbers but of the quality of volunteers that the application system was producing: "indeed, after all it is impossible not to agree, that only a young energetic person, absorbed by interest in aeronautics, can completely devote himself to his service, and not one who is bored, trying to amuse himself in the noise of life in the capital, by chance presented with the opportunity to get into the Officers' Aeronautics School." The author summed up his critique by pointedly contrasting the understanding and appreciation of the needs of military aviation between the civilian and military communities. Unlike the citizens throughout the Empire who energetically and patriotically supported aviation though donations and sympathy, "many unit commanders and their higher commanders, because of petty personal benefit, because of the fear of falling into the disfavor of their immediate commander, not only do not feel it their sacred duty to in every way accommodate the good impulses of their subordinates, but it seems do not share the general interest in the arising power of our Motherland."[17]

The fears expressed in this article were a bit overwrought, while various factual inaccuracies in terminology and details attest to the author's own imperfect understanding of the application system for aviation service. Yet, like the criticisms raised by Dukhanin, and like the complaints regarding the alleged resistance of engineers to the opening of aviation to all branches of arms, the accuracy of such charges was in a sense less significant than their appearance. For perception was a concern as the senior command attempted to establish and legitimate aviation service to an all-army constituency, not least to the junior officer and potential volunteer who might be reading *Razvedchik*. These critiques drew a quite negative picture for such officers, one that very much resembled the portrait painted later by Soviet historians. They railed against the obstacles faced by their ideal volunteer—a young, serious aviation enthusiast. Easiest to indict was the culture of inertia and petty self-interest which blinded many officers to the importance of aviation. These critiques went further, however, and addressed the application criteria themselves, the command's own willingness and ability to enforce these criteria, and, perhaps most subversively, the very right and authority of the army leadership to determine the requirements of aviation service. Thus, the policy for volunteers did not provide for "a careful and strict" selection, did not adequately measure commitment or aptitude for aviation service, and gave too small a voice to those truly qualified to assess such qualities, the pilot instructors.

Rather than the necessary elite of the talented and devoted, of which there were many languishing in their line units, it was instead young officers seeking to amuse themselves "in the noise of the life of the capital" who were able to sign on for pilot training as a fashionable trend. As this implied, such officers were not even subject to the established application process, imperfect though it was. Memoirs and Soviet historiography later made such allegations plain and observed that a well-placed patron or some similar connection was necessary for posting to an aviation school. Although the anonymous author in *Razvedchik* specifically mentioned the Officers' Aeronautics School, it was the Sevastopol Aviation School which was more often the recipient of such criticism. Despite the uniform application criteria and the ostensible control over the admissions process by the Engineering Administration and later the Aeronautics Unit, it was believed that the Grand Duke and his Department of the Air Force managed to ensure that "their" types of officers, measured by proper social background and prestige of unit, were selected and admitted to the Sevastopol School. In contrast to the alleged predilection for engineers which prevailed at the OVS, this meant primarily providing for the "golden youth" of aristocratic families, and thereby ensuring a greater percentage of fashionable playboys and dilettante aviators.[18]

Official reports, unofficial criticism, and lingering gossip all testified to the tangible difficulties, divided counsels, and diverse forms of resistance which attended the efforts to recruit a cadre of officers for aviation service. In light of all this, what types of officers did indeed volunteer for pilot training? To what extent did they fit the ideal provided for in the regulations, and to what extent did the application process and its outcomes match the goals of official policies? Suggestive are the formal petitions submitted by officers wishing to volunteer, many of which were accompanied by the recommendations required on the part of the applicant's commanding officer. In a letter of Nov. 17, 1910, the commander of the Ivangorod fortress petitioned Gen. Aleksandrov, commander of the Engineering Administration, to render assistance to a subordinate wishing to be admitted to the Aviation Section.[19] He observed that the officer in question, Second Lieut. Kushke,

> is a sober, able, brave, energetic officer, loved by his comrades, a good bicyclist, [who] has carried out free flights and is so experienced in conducting observation (he belongs to the observation team) that he drew the attention of His Imperial Highness the General-Inspector of Artillery, who even remembered his family name, and by his character obviously will not limit himself to flights on balloons, but, having received the appropriate training, will take the most active part in flights on airplanes.[20]

In a secret letter, dated Aug. 28, 1913, the commander of the 33rd infantry regiment appealed to Capt. Fogel', adjutant of the Grand Duke, on behalf of a Second Lieut. Kostechkin who wished to enter the Sevastopol Aviation School. The commander noted that "I consider it my duty" to add that Kostechkin

> is a completely fine and conscientious officer. He treats his affairs with complete attention and I hope that he will treat aviation with no less effort than his line unit, the more so as, having a strong penchant for it, he is seeking to be enrolled in one of the

> the aviation schools. At the present time of maneuvers he has been assigned to the 3rd aviation company to train himself for the role of officer-observer.[21]

The commander added that while Lieut. Kostechkin had received a reproach in his attestation for lack of discipline, "this is merely due to his youth, and as far as I have noticed, these defects have completely disappeared." Another letter to Capt. Fogel', dated July 15, 1913, came from the commander of the 147th infantry regiment. It concerned a request of one of his officers, Second Lieut. Padosek, to be assigned to the Sevastopol School. Noting that he was forwarding a copy of the required attestation, the commander added that "I am informing you that this officer proved to be in the regiment impeccable in both service and moral aspects, and I, not hindering him in his aspiration to become a pilot, can only regret that, in light of the attention you have given him, he will probably soon leave the ranks of the regiment entrusted to me."[22]

There is indeed evidence that patronage, official and unofficial, attempted and successful, played its part as well. Such was the case with Staff-Capt. Al'mkvist of the 7th Kronstadt fortress artillery company. In a report to his company commander on Sept. 28, 1910, he noted that his original petition to be assigned to an aviation school had not been passed on.[23] He added, however, that on Sept. 22, at the aerodrome of the "Wing" Association, the Main Commander of Kronstadt, General-Lieut. Artamanov, "having found out about my ardent desire to study aeronautics, promised me full assistance" in getting into an aviation school, and had ordered him to inform the regimental commander.[24] In a subsequent letter to the Chief of Staff of the Petersburg military district, General Staff General-Lieut. Artamanov noted that "personally knowing Staff-Captain Al'mkvist and his passion for aviation" he was petitioning the Commander of the Petersburg military district for the assignment of Al'mkvist to an aviation school.[25]

These petitions and recommendations provide some insight into the volunteer impulse and the elusive role of patronage, prerogative, and passive resistance. Many of those who answered the call to take to the skies did indeed share "a passion for aviation," had already taken steps to train and prepare themselves for such service, and had exhausted all avenues to ensure their acceptance for pilot training. Their initiative and enthusiasm was palpable, sometimes to a fault; Lieut. Kostechkin, for example, had been cited for his lack of discipline and his need for "firm supervision." In turn, the recommendations illustrated that not all tsarist senior officers viewed narrow self-interest as the raison d'etre of their service. Many were happy to satisfy the wishes of their junior officers and assist the development of aviation through such a valuable contribution. At the same time, despite such evidence of official solicitude, various petitions submitted by officers requested assistance or intervention because they could not prevail upon their commanders to release or recommend them. Several of the recommendations cited above came after initial efforts, through the usual channels, were refused. Further, we do not know how many petitions were peremptorily refused, dying at the level of local administration, and thus never entering the documentary record. Moreover, confirming the suspicions raised in the anonymous *Razvedchik* piece, various unit commanders submitted reports to the Aeronautics Unit of the General Staff in which they complained of staffing shortages and/or requested the return of their officers from aviation service, while in their letters various volunteers cited these same concerns as the reason their initial petitions been denied.[26] Lastly, allegedly acting as a powerful yet amorphous filter for the entire application process was the social bias on the part of the Department of the Air Force, which of course was not disposed to trumpet or publish its alleged informal measures of exclusion.

Evidence of the application process itself is thus still fragmentary; more clues as to what it wrought can be found by looking briefly at the graduation lists of the respective schools. Of a list of 28 officers finishing the course of the Aviation Section of the Officers' Aeronautics School in 1912 and receiving the designation *voennyi letchik*, sixteen were members of aeronautics units, seven of sapper units, three of mining units, one of a pontoon unit, and only one was from a non-engineering line unit. Added to this list were those fifteen officers who had been trained as pilots before the official formation of the Aviation Section, either abroad or at Gatchina in 1910 or under the Temporary Aviation Section in 1911.[27] Again, this group, which comprised most of the permanent command and training staff of the Aviation Section, was made up overwhelmingly of officers from existing aeronautical units. By 1913, the composition of the graduating class of the Aviation Section of the OVS had significantly changed, reflecting the admission of officers from all branches of arms. Of this group of 28 officers, nine were from aeronautical units, eight were from line infantry and rifle units, four were from artillery brigades, three were from grenadier regiments, and there was one each from a dragoon regiment, from sapper and mining units, and from the corps of border guards.[28]

In contrast, a list of the permanent and temporary personnel of the Sevastopol Aviation School in April 1911 listed Colonel Odintsov, Capt. Kedrin, and Lieut. Dybovksii (the latter two from the navy) as the permanent staff; Lieut. Piotrovskii, Staff-Capt. Matievich-Matseevich, and Lieut. Rudnev as officer-supervisors; civilian supervisor M. Efimov; and 23 officers in training. Of these, eight were from aeronautical units; six were from Guards units; two were from the Fanagoriiskii grenadier regiment; one was from the 6th Hussar regiment; three were from line infantry and artillery units; and three were from naval units.[29] While not grossly disproportionate, the Guards units were nevertheless very well represented. Two lists of officers who had finished the Sevastopol School and had received the designation *voennyi letchik* or who had satisfied all conditions for it, dated February 1913 and October 1913 respectively, present a different picture.[30] The first list of fourteen officers included seven from line infantry and rifle regiments, three from artillery brigades, one each from a sapper and railroad battalion, and one each from a Guards artillery brigade and a Dragoon regiment. The second list of 24 officers included twelve from line infantry and rifle regiments, six from various artillery units, one each from an aeronautical and railroad battalion, one from the 1st Guards artillery brigade, one each from a grenadier and hussar regiment, and a sotnik from the Siberian Cossack regiment. The permanent staff of the school included a prince, Muruzi, as its commander and Staff-Capt. Viktor-Berchenko of the Preobrazhenskii Guards regiment as supervising instructor, but also included the civilian Efimov.

The lists of graduates from both aviation schools, which indicated only rank, branch of service, and unit, give only a bare description of the cadre of officer aviators. Because individual service records for the most part were not preserved with these lists, the social composition and the closely related issue of the educational backgrounds of these officers are more elusive to determine. From an impressionistic and anecdotal survey of service lists that are preserved in the records, however, some initial observations can be made: both nobles and non-nobles were well-represented amongst the trainees of both schools; for primary schooling, the cadet corps and civilian *real* schools and *gimnazii* were about equally represented among these officers; for secondary and higher learning, the great majority had attended military *uchilishche* and subsequently the specialty military academies, though there was a smattering with civilian university or higher technical education as well. As the composition of the aviation services changed from one heavily represented by engineers to a more diverse composition, including many from the infantry, the number who had attended the academies dropped.[31]

Both the virtues and failings of the recruitment system were in part illustrated by the experience of one of our protagonists, Lieut. Petr Nesterov of the 9th Siberian Artillery brigade. Despite his active interest in aviation, and despite serving in an aeronautical unit, his ambitions were initially frustrated.[32] Only "with great difficulty, owing to the absence of connections and high acquaintances" was he accepted not to the Aviation Section as he petitioned, but to the OVS itself; even then the first doctor to examine him for the required medical certification declared him unfit.[33] Nevertheless, accepted he was, and due to his expansive knowledge in the sphere of aeronautics he quickly passed the regular course of the OVS in the first rank and requested to be transferred to the Aviation Section.[34] Gen. Kovan'ko duly submitted a petition on Nesterov's behalf. Chief of the General Staff Zhilinskii's report to the War Minister on the matter noted Nesterov's "great interest in aviation" and his experimental work in the field. This led Gen. Kovan'ko to grant Nesterov special permission to train on airplanes, as a result of which he had already passed the examination for *voennyi letchik.* Zhilinskii thus recommended the approval of the transfer, a request granted by War Minister Sukhomlinov.[35] It was Nesterov who was the sole non-engineering officer to graduate from the Aviation Section of the OVS in 1912.

An officer cadre drawn from all branches of the army and attracting the best, brightest, and most willing: from 1911 onward this was the official ideal for the budding aviation service and its training institutions. While real and perceived difficulties persisted, bare statistical records demonstrate that this vision was largely realized. Aviation service was rapidly transformed from the preserve of highly trained technical specialists to a more educationally democratic and diverse community of officers. Both stories in the contemporary press and subsequent memoirs attested to the spectacle of seeing the uniforms of the infantry, the cavalry, the artillery, the engineers, and the Guards milling about together at the aerodromes of Sevastopol and Gatchina. In a cultural sense as well, the decision to accept volunteers from all branches of arms meant that aviation would rapidly outgrow the parenthood of the engineering and aeronautical services. This had the effect of decreeing that a culture of aviation service would have to engage and hopefully transcend the internal divisions that the Imperial army was notorious for. It was thus their actual training and experience as aviators that would or would not truly bind such officers together as a cadre of *voennye letchiki.*

Training

Policies of recruitment were only one piece of the aviation puzzle. Indeed, they were merely a necessary prelude to the challenge of training and utilizing the personnel of the new air arm. Here the focus would shift from high policy to the setting of the aviation schools themselves, which in turn did far more than simply train new pilots. At both the Aviation Section of the Officers' Aeronautics School and the Sevastopol Aviation School, efforts continued to refine piloting techniques, to improve technical equipment, and simply to learn more about the science of aerodynamics and the experience of flight. In this regard over time certain technical standards were established for aircraft, rules for aerodrome operations were codified, training procedures were to an extent regularized, and test examinations for the designation of *voennyi letchik* were standardized.[36] Due to the cosmopolitan nature of aviation development, such efforts were again aided both by domestic experience and information gleaned from abroad.

However, there was still far from perfect agreement regarding which skills would be required by the new cadre of military pilots and how best to impart those skills. There were instead continual debates about the exact nature of the role and function of a pilot and his soon to be companion, the observer. This owed in large part to competing ideas about the basic nature of flying itself, a controversy endemic to the early aviation experience and faced by all of the nascent European air forces. Central to this debate was the question of whether flying was an art, a science, a skill, or some combination of each.[37] The view of flying as art stressed instinct and innate talent on the part of the pilot; the conception of aviation as science necessarily implied intellectual preparation and rigorous study of the subject. Viewing flying as a skill in a sense straddled these two extremes; piloting was something that needed to be learned, yet largely through practical training and experience, without implying any exclusivity of education or intellect.

Depending upon which definition was chosen, one made certain assumptions about not only the proper scope of knowledge and training required for a pilot, but also about the types of mental and emotional qualities necessary for him to possess. In both the European and Russian aviation services, these competing ideas were manifested most vividly in the dichotomy of what might be called "pure flier" vs. complete aviator. In the first group were those officers who considered their role and function to be merely that of flying the airplane. The maintenance of the aircraft, and indeed any extensive knowledge of the technical aspects of the aircraft and its engine, were solely the responsibility of the soldier-mechanic assigned to each officer. The "white hands" [*beloruchki*] considered this latter type of work beneath them. Instead, their purpose was to fly, a claim often based upon a demonstrated talent and natural instinct to pilot an airplane.[38]

Opposed to this group were those who argued forcefully that a pilot had to be thoroughly acquainted with all technical aspects of the airplane and its equipment. In fact, rather than rely solely on his mechanic, an aviator had to inspect both the condition of the plane and the work of his mechanic before each flight, drawing upon a body of technical knowledge at least as extensive as the mechanic himself. Among the most eloquent proponents of this view was none other than Mikhail Efimov, whose training and interest in both the technology and theory of flight was largely unparalleled. Efimov absolutely insisted on thorough technical training as well as a good understanding of aerodynamics for pilots. As he succinctly put it, "aviation is a very serious sport," one which required "meticulousness, circumspection, composure"; he stressed that "a person is not born an aviator; aviation must be learned." In his view, getting used to the experience of flight came only slowly and with difficulty; only after about one hundred hours in the air did a person come to pilot his machine almost instinctively.[39] Such an attitude, needless to say, precluded a spurious or casual approach to piloting and demanded a thorough knowledge of one's craft. The same view was held by Lieut. Petr Nesterov, who again, even before being trained as a pilot, had immersed himself in the study of aerodynamic theory and had even designed experimental aircraft.

The process of reconciling these conflicting standards was again played out within the context of the different philosophical approaches toward military aviation among the Sevastopol Aviation School, the Officers' Aeronautics School, and the Aeronautics Unit of the General Staff. Despite the rapid expansion of aviation in this period, the permanent staffs of these organizations remained quite stable and continued to serve as repositories and conduits of their respective traditions, however young.[40] These competing philosophies of training and mission were manifested in an emphasis upon different priorities and skills; these might roughly be seen as the difference between training aviators and producing fliers, between comprehensive instruction and targeted, rapid training. Such differences became even more important in the Russian context. As noted, unlike France and Great Britain, the Russian army for the most part could not and did not wish to rely primarily upon private aviation schools to provide preliminary pilot training for officers. Conducting the bulk of training themselves, the two military aviation schools thus provided notably different experiences of socialization and acculturation for the classes of future *voennye letchiki*.

The instructors of the Officers' Aeronautics School approached their task largely from the standpoint of technical specialists. Although aviation was rapidly outgrowing the parenthood of the engineering service, the longer history and established institutional culture of the OVS continued to exercise its influence on the staff of the Aviation Section. Accustomed to a broader and deeper perspective on issues of technological innovation, the staff of the Section stressed a rational over intuitive approach to pilot training, rigorous study, and thorough preparation in technical matters and the theory of aerodynamics. To recall an incident discussed earlier, even in his denials of charges that he chose to thwart the entrance of non-engineering officers into the OVS, General Kovan'ko's concern over the technical qualifications of such officers was palpable. Memoirs further attest to the high level of technical and theoretical instruction for Aviation Section trainees,[41] while popular histories note that the rigor of its training made the Section the "more prestigious" of the two aviation schools.[42] Even Soviet histories recognize the quality of the training provided there.[43]

In fact, attention to issues of theory and technology was so great that upon external review the curriculum and training program of the Aviation Section were revised in this period to include more practical flight instruction. In further recognition of the enduring parentage of the OVS, in Gen. Shishkevich's master plan of organization for aviation the General Staff envisioned transforming the Aviation Section into an independent aviation school in the near future and ultimately relocating it to Moscow.[44] In turn, "in view of the fact that at the present time the school is strongly overburdened with theoretical courses in aeronautics and sciences related to it," the report recommended removing some of these courses and thereby giving the Officers' Aeronautics School the opportunity to increase practice on flights in dirigibles, which at present were conducted on an "insufficiently broad scale." In light of such an environment, with the exception of Mikhail Efimov, the major proponents of the thorough technical and theoretical preparation of pilots—officers such as Col. Ulianin, Lieut. Kovan'ko, and in particular Lieut., later Staff-Captain, Nesterov—were all instructors at or products of the Aviation Section of the OVS. Some of these officers had previously been engineers and aeronauts, some had come to aviation from other branches of arms, but all of them, in addition to their activity as pilots and instructors, also designed and experimented with their own aircraft and related equipment.[45]

In contrast, imbued with the heady sense of urgency and mission imparted by their royal patron, the staff of the Sevastopol Aviation School approached their task as the vanguard of a future mass of pilots. As attested to by the public proclamations issued by the Grand Duke and his Department of the Air Force, providing the army with a large cadre of aviators as rapidly as possible was the raison d'etre of the Sevastopol School's existence. This entailed an overriding emphasis upon practical flight instruction, not intellectual pursuits. While preparatory theoretical classes in aeronautics at the St. Petersburg Polytechnical Institute became an early feature of the Sevastopol curriculum, these classes were included only to enhance pilot training.[46] In the report of its activity for 1913, the OVF emphasized that upon taking command of the Sevastopol School in 1911 Col. Odintsov had received an order from the Grand Duke that he was "to be guided in his activity by the fundamental demands of the then newly worked out regulations, and primarily to keep in mind that *the ultimate goal of the school's existence is to produce military pilots,* that is those who, possessing to the highest degree *the skill to fly,* would simultaneously possess *the skill to observe* from the heights that which is required for military operations"[47] The report added that from this time on training began to take a "more correct" course and that the school "little by little received a purely military direction."[48] This oblique acknowledgment of the somewhat amateurish origins of the Sevastopol School was made more directly in an earlier report of the school's activity upon the occasion of its one year anniversary on Nov. 19, 1911, which noted that in March 1911 the Grand Duke had recognized the need to reorganize the school, "to take as a starting point strict organization and military education of future military aviators." It was then that the regulations had been revised and Col. Odintsov had been placed in command; moreover, Odintsov was given the rights of a regimental commander in terms of disciplinary authority over officers posted to the school.[49]

Such differences in approaches and philosophies between the two schools did not remain merely symbolic ones, as beginning in 1912 pilots completing the course at the Sevastopol Aviation School were assigned to the Officers' Aeronautics School or existing aviation squadrons for a period of supplementary training, followed by their final examination for the designation of *voennyi letchik*. Despite the common training requirements and criteria among the two schools, there were at times allusions to the somewhat less than satisfactory technical preparation and knowledge of Sevastopol pilots. Notable in this regard was General Shishkevich's sarcastic remark to Capt. Fogel' of the Department of the Air Force that pilots trained at the Aviation Section of the OVS, as opposed to the Sevastopol School, did not forget how to fly over the winter.[50] One could take this, as Capt. Fogel' indeed did, as a swipe at the abilities of Sevastopol trainees in general. A more concrete example came in a secret report of Colonel Ulianin, commander of the Aviation Section of the OVS, to General Kovan'ko in March 1913, in which he requested the dismissal from the school of two officers, Lieutenants Treksin and Beliakov, trained at the Sevastopol Aviation School.[51] Ulianin reported that "during their stay at the Aviation Section their small aptitude for flying and extremely poor technical preparation became clear." In the attached attestations of 1912 for these officers from the Department of the Air force, the comments concerning Lieut. Treksin noted that he "has no aptitude for flying. Does not like to fly. Excellent, industrious officer. Very efficient."[52]

Snide remarks and occasional expulsions bore witness to the differing training philosophies of the two schools; harsher evidence was provided by the frequent crashes, often fatal, that marked the early years of aviation. Following such fatal mishaps, it was standard practice to appoint an investigative commission to delve into the reasons for the crash. Due to the relative novelty of aviation, the investigations often were unable to correctly determine an exact cause.[53] Yet the commissions often discovered the failure of the deceased officers to inspect or properly prepare their planes before flight. If they had taken the time to do so, they might have detected and repaired the defects believed to be responsible for the accident. Likewise, when asked about the numerous aviation crashes both at home and abroad, Mikhail Efimov argued that the primary cause was most often lack of technical training; he especially cited the *beloruchki* who disdained technical matters and believed it was their duty only to fly.[54]

Yet despite the fact that Efimov held the post of instructor there, crashes, especially those resulting from seemingly preventable causes, were particularly a problem for the Sevastopol Aviation School. A newspaper item of 1912 noted that six officers had died during the past year at Sevastopol; the final tally included ten officers killed at the Sevastopol School itself from 1910–1914, plus at least another five who were trained there and subsequently crashed and died while on duty. In contrast, there were four officer fatalities at the Gatchina school itself, plus a few more subsequently of Gatchina trained officers.[55] Naval Lieut. Iatsuk, chief instructor of the aviation school of the Imperial All-Russian Aeroclub, drew the requisite conclusions from this disparity by early 1912. Noting the achievements of Russian military aviators in 1911, he observed that

> they prove that in the Russian army too the airplane can become equal to other instruments of war. The typical distinction of the correctness of the organization of the training of military pilots is the almost complete absence of crashes during training flights at the Aviation Section of the Officers' Aeronautics School. In the school of the Department of the Air Force, on the contrary, there were many crashes.[56]

In part because of such differences regarding philosophies and methods, despite efforts to standardize curricula and instruction regimens the quality of pilot training, and the resulting quality of the military aviator cadre as a whole, remained uneven and inadequate.[57]

Death certainly provided persuasive if jarring testimony in the debate over the proper scope of training and competence for new pilots. In addition to such disagreements, however, the nature of their role and function was further defined around the juxtaposition of pilots and observers. As the airplane was initially viewed not as a weapon but as a vehicle for reconnaissance, the role of observer was crucial; it would become even more vital once artillery spotting became a primary function of aerial operations. The Russian air services were already developing such tactics as early as 1913, led again by the visionary Lieut. Nesterov. With his initiative and ideas as a guide, the 3rd aviation company conducted practical exercises in artillery spotting in conjunction with units of the Kiev military district in the maneuvers of summer 1913.[58]

In theory at least it was relatively quickly accepted that observers should ideally be officers of the General Staff, formally acquainted as they were with matters of topography and reconnaissance.[59] This decision too was based in part upon the French experience. A presentation of GUGSh to the War Council on April 4, 1911 summed up the matter succinctly. It observed that "the fundamental task of the air force is close and long-range reconnaissance for military purposes. In order to be able to use this new and powerful means of reconnaissance with full understanding, it is necessary to train for this a sufficient number of officers of the General Staff."[60] The presentation thus proposed to establish a system of assigning such officers, ten per year, for a period of 3–4 months of practical training at the Officers' Aeronautics School, according to a program worked out jointly by the General Staff and the Main Engineering Administration.[61]

A question that received a good deal of attention in this regard was whether such General Staff officers should also be trained as pilots. Some interesting observations were provided by the General Staff representative to the OVS commission established to evaluate the results of the first year of such training.[62] In his report, dated Sept. 2, 1911, General Staff Colonel Bulgarin noted the general success of the training program, but nevertheless pointed out that of all these General Staff officers not one had been trained for independent flights. Thus, it was "inecessary to resolve the fundamental question- should officers of the General Staff be in a condition to fly independently or is it sufficient to develop in them only those qualities necessary for a good observer." He observed that

> it turns out to be necessary to take into account a natural psychological phenomenon: to the extent that an officer becomes familiar with the piloting of the airplane, his trust in the pilot will go down, and the better he himself pilots, the more difficult it will be for him to remain in the role of a passenger observer. Such renowned pilots as Efimov and Rudnev absolutely refuse to be in the role of observer. In light of these considerations, all those present at the examinations expressed the opinion that the training of an observer in independent flight is not only unnecessary, but might have a negative effect upon his training as an observer.[63]

He recommended further that "the production itself of observers will be significantly speeded up if General Staff officers receive instructions not only from officer aeronauts, who in the majority of cases regard flights from a purely technical and too narrow point of view, but from their comrade officers of the General Staff who have already become practiced observers."[64] After making some more detailed recommendations, Bulgarin observed that

> with the question put thus it would be natural to refuse to train officers of the General

> Staff in independent flight... Moreover, the time is not far off for the use of airplanes with combat purposes, and perhaps even as entire aerial squadrons, and now is already the time to think about the training of suitable commanding personnel. If it is recognized as necessary, besides officer-observers, to form at least a small cadre of officer-pilots, then it will turn out to be necessary to assign such officers for the full course of the aeronautics school. A small cadre of officers of the General Staff who are able to fly independently will serve as an excellent example for other arms of service and as a leader in the application of aeronautics to military affairs.[65]

Colonel Bulgarin concluded his report by noting that "everything stated in the present report is not the result merely of my own personal beliefs. These opinions are also shared by General-Lieut. Pavlov, those officers of the aeronautics school, headed by Major General Kovan'ko, who were at the examinations, and finally by those officers of the General Staff who were assigned to the school."[66] Bulgarin's conclusions were approved by the examining commission and adopted as policy for following years.

An alternative view was provided by General Staff Colonel Agapeev, who in a welter of correspondence argued with the Engineering Administration and the Officers' Aeronautics School for the right to be trained to fly.[67] In a report of July 11, 1912, Col. Agapeev asked that Gen. Kovan'ko provide his own conclusions and then forward the attached memorandum to the General Staff. Noting that he had been assigned for the second time to the OVS for advanced training in aerial observation, Agapeev requested pilot training: "I consider such study necessary in order that, in light of the impending transfer of the matter of aeronautics to the jurisdiction of GUGSh, there will be in the corps of officers of the General Staff officers who have studied the matter of aeronautics and aviation in sufficient detail." Agapeev noted that while several General Staff officers were trained to fly independently on balloons, not one was trained on airplanes; both Colonel Odintsov and Prince Muruzi, he stated, could not fly independently.[68] He added that he wished "to prove the fallaciousness of the view" that training an observer to fly would be detrimental, and reiterated the broader motivations that underlay his request: if General Staff officers were trained as pilots as well as observers, "interest in aeronautics and aviation among officers of the General Staff will beyond any doubt grow. The weighty significance of aeronautics and aviation in military affairs has been firmly established, and insistently points to the necessity of a more serious study of this affair by officers of the General Staff."[69] In a subsequent report submitted to the Main Engineering Administration,[70] Gen. Kovan'ko reiterated Col. Bulgarin's conclusions, noting that this question had been decided by the examining commission the previous year with the approval of the Engineering Administration, and added that "fully sharing the opinion of the commission, I do not find grounds for changing the procedure."

The reply of the Engineering Administration agreed. Not to be deterred, Agapeev submitted another letter to Kovan'ko. Complaining that his memo had not been passed on to the General Staff as he requested, but had instead been decided by the Engineering Administration, he emphasized again that "in light of the impending transfer of the Aeronautics Unit to GUGSh the question of training officers of the General Staff in piloting should doubtless be decided by this administration."[71] Despite his persistence, a subsequent message from the Aeronautics Unit requested that Agapeev and his colleague General Staff Capt. Kovalevskii be sent back to their permanent posts.

The debates, reports, and statistics cited above provide merely a snapshot of the daily and ongoing task of training a cadre of Russian military aviators in the years 1910–1914. Yet the cleavages within this new community were readily apparent. The contrast between aviation as art and science, between the concept of pure flier and that of complete aviator, provided competing sets of standards of what it meant to join the fraternity of pilots. One set stressed technical education, objective standards of competence, and acquired skill. The other was based to a greater extent upon a more caste-like selectivity which emphasized innate talent; it too, nevertheless, focused upon functional specialization. These competing ideals regarding the nature of aviation and the proper definition of a pilot's competence would have a marked influence upon the elaboration of a professional identity for the new *voennye letchiki*.

This process was to a certain extent complicated by the role of the General Staff. The stated rationale for its assumption of overall jurisdiction over aviation was the claim that it was the one institution whose concerns and competence embraced the needs and mission of the army as a whole. Noteworthy was the almost seamless way in which such thinking was applied to the specifics of flying. Colonel Bulgarin reiterated the belief of the higher echelons that pilots, by training and inclination, would view aviation almost exclusively from a "purely technical and too narrow" point of view. Thus, while General Staff officers need only be trained as observers at present, as the airplane soon became a serious weapon, it would need such officers to provide overall leadership and guidance, and serve as "an excellent example" to the other branches of arms. Col. Agapeev cited exactly the same motivations as he lobbied for the right to be trained to fly. Both of these officers implied that the aviation service itself was incapable of producing its own cadre of officers who would truly be able to appreciate the significance of military aviation.

Although they agreed that General Staff officers need only be trained as observers, General Kovan'ko and his OVS officers no doubt chafed at the reluctance of General Staff officers to cede a leading role to the pilots themselves. This underscored the continuing struggle to define competence and apportion authority in the realm of military aviation. In contrast to the General Staff, a competing perspective regarding both recruitment and training was to allow the greatest latitude and discretion in all aviation matters not to doctors or commissions or engineers or bureaucrats or the General Staff, but to the real experts, the aviators of the line. Yet, as seen in their varying motivations to volunteer and the differing training experiences they received, the aviators themselves did not speak with a unified voice. Tensions that revolved around issues of competence, prerogative, and professionalism would only become sharper as the newly ordained *voennye letchiki* left the training schools to pursue their careers in the aviation squadrons.

Authority, Leadership, and Expertise

By bringing together the graduates of the different schools, the setting of the aviation squadrons emphasized the differing interpretations of the duties of a military pilot, as well as what some would argue was the related distinction between serious aviators and playboy flyers. At the same time, the new environment of the squadrons at least provided the opportunity to reconcile these competing standards and ideals. The larger problems of defining expertise and exercising authority in the realm of military aviation thus took on both familiar and novel forms as the *voennye letchiki* performed their daily duties and tended to their flying.[72]

Such tensions were exacerbated by the generational rifts—it would be fair to deem them such in the rapidly evolving world of aviation—within the community of military pilots. The explosive growth of the aviation service must constantly be kept in mind in discussion of these issues. In the fall of 1910 less than ten officers represented the sum total of Imperial Russia's cadre of military aviators. By 1913, Gen. Shishkevich was envisioning a force of 64 aviation squadrons with four officers per squadron, while in the same year the OVS and the Sevastopol Aviation School produced approximately 70 new pilots. Due to this extremely rapid expansion, what had originally been little more than a close-knit and relatively homogeneous club, especially at the OVS, was being transformed into a branch of service larger than any of its members. While the commanders and staffs of the aviation schools functioned as powerful sources as socialization and acculturation to aviation service, the relatively short term of the courses (less than a year), the numbers and variety of trainees involved, and the very novelty of the aviation institutions themselves helped to ensure that graduating officers could continue to fashion their own unique ideals of service. As again was common to the military aviation experience throughout Europe, not all of these officers placed unselfish and unstinting devotion to aviation at the top of their priorities.

Thus, for those officers who had energetically devoted their energies to the successful development of aviation, frustration, disgust, even outrage was the response to their comrades who flew the minimum hours per month to garner their flight pay, who took no interest in experimentation or the newest advances in aviation technology, who read none of the aeronautical publications, and who demonstrated at best indifference, at worst fear, to the actual business of flying. The preferred cliché of the literature usually weaves together several stock elements of this "unprofessional" ideal and casts such officers as members of the Guards, usually nobility by birth, and most often products of the Sevastopol School.[73] The presence of such dubious *voennye letchiki*, and their ability to flout the spirit if not the letter of the regulations and adhere to their own definition of duty, did not go unnoticed by their fellow aviators or the command itself. Again, men like Mikhail Efimov, Col. Ulianin, and Lieut. Iatsuk of the All-Russian Aeroclub consistently proselytized in the aeronautical and popular press for a thoroughly professional, technically and theoretically informed ideal of service. Col. Veigelin regularly published *Vozdushnyi spravochnik*, a yearly compendium of technical developments in aviation, while in 1912 the 7th aviation company published the pamphlet *The Airplane and its Combat Significance*, based upon a report delivered in the officers' club by Lieutenant Mikos. The pamphlet provided detailed analysis of issues of aerodynamics, technical capabilities of aircraft, and the application of aviation to various military tasks.[74] The belief that thorough technical knowledge was fundamentally integral to success in the air were expressed even more forcefully in an order of the day issued to the 3rd aviation company following Staff-Captain Petr Nesterov's famous flight from Kiev to Gatchina in June 1914. Praising Nesterov's achievement, the order noted that it hoped the example provided by the flight

> will stir up competition among the other officers of the 3rd aviation company and instill in them the realization that for the success of the affair not only is daring needed, but also a thorough understanding and precise knowledge of the properties of the airplane and motor. Staff-Captain Nesterov is strong in these and is obliged to them for his successes. Whoever does not do this or does not persistently enough lead their subordinates along this path is morally responsible for accidents with pilots and observers.[75]

For advocates of this vision of professionalism, more enduring and serious difficulties were posed by the problematic relationship between seniority, rank, and authority within the aviation services. If this was a problem among those officers at the echelons of leadership, it was more so at the level of the line, as aviation developed so rapidly that often yesterday's pioneer was to an extent today's relic. It was indeed newer pilots, armed with better technology, accumulated experience, and more advanced techniques, who were thus in a prime position to innovate, at times even outpacing their instructors. In turn, it was yesterday's pioneers who were already advanced in rank, often in positions of administration where flying was no longer necessary. To again take the example of Mikhail Efimov, he had been trained to fly in 1909, where maintaining stability in flight and avoiding undue maneuvers had been the key to living a long and prosperous life as an aviator. For this reason Efimov had initially argued against overly sharp turns and severe banking by the aircraft. Yet by 1914 both the theory and technology of flight had advanced far enough that such maneuvering was not only recognized as possible but necessary for military aviators. Efimov himself, who continuously innovated and adapted over the course of his career, indeed helped develop some of these new methods and techniques. Yet not all pilots were so flexible or adaptable.

Of course, even when certain innovations were initially resisted by established pilots, these officers generally realized that developing and assimilating new piloting techniques ultimately benefitted all within the fraternity of aviators. The tensions between seniority, rank, and authority were thus much more striking and potentially disruptive when they pitted the cadre of line aviators against those in positions of command. The phenomenon of creeping obsolescence in the higher ranks was not wholly unique to aviation or the Russian army. It is a problem that all armies, particularly their technically oriented units, must continually grapple with. Yet in the tsarist aviation services these problems again manifested themselves in particularly accentuated form. Thus, if the commanders of the aviation institutions often vigorously asserted their competence and qualifications, it is ironic that two of them—Generals Kovan'ko and Shishkevich—were not trained aviators. Kovan'ko had been an engineer and balloonist, while Gen. Shishkevich again was merely an aviation enthusiast who possessed the requisite rank and General Staff qualifications. Colonels Odintsov and Muruzi had some flying experience and training, but again, at least according to Col. Agapeev, could not fly independently. In contrast, Col. Ulianin and the supervisors and senior staffs of both aviation schools were all trained pilots. This was indeed a crucial distinction. This state of affairs whereby aviators had to defer to non-aviators and accept orders which could influence not only their careers but their very lives was a source of potential resistance at the level of the line. Again, the most striking example was the refusal of a number of officers from the first class of pilots trained at the Sevastopol School to take postings in the aviation squadrons. To recall Col. Odintsov's report on this matter, it seems he almost relished the opportunity to inform the War Minister that this decision was the result of these officers' belief that aviation could not successful conclusion while under the jurisdiction of the engineering service.[76] While these officers' greater faith in the non-aviators of the General Staff was to be somewhat disappointed, the incident was powerful assertion of the desire and ability of military aviators to construct their own model of duty and service.

Direct or dramatic testimony on the part of line aviators was rare, however. More often, clues were provided by those who claimed to speak for them. The civilian aviator N. E. Popov again eloquently expressed the perceived antagonism between aviators and engineers by stating that "eagles cannot be subordinated to moles!"[77] In a similar vein, an editorial commentary in *Sevastopol Illustrated Aviation Journal* in February 1911 observed that a proposed program of examinations for the designation *voennyi letchik* was being worked out both by a special commission of the General Staff in St. Petersburg and by the Council of the Sevastopol Aviation School.[78] While not denying that the members of the Petersburg commission were probably interested in aeronautics, the editors noted that the commission should be made up of people who had actually risked their lives flying, and all of those were in Sevastopol. The editors added that details in the press regarding the work of the Petersburg commission, as well as the individual opinions of several instructors, had further convinced them that such decisions should be made by the experts in the matter.

Such generational and professional rifts were perhaps sharpest between the aviators and their close cousins, the aeronauts. The relative merits of various aeronautical technologies were still being openly debated throughout this period, but astute observers realized that the airplane, in terms of the scope of its potential, was clearly poised to supplant balloons and dirigibles. In his memorandum of November 1911 to the Emperor, War Minister Sukhomlinov had cited as one of the "fundamental" tenets that the War Department, while not totally rejecting the latter technologies, "should now direct all of its efforts to the soonest possible supplying of the army with airplanes."[79] Yet in many ways aviation continued to be subordinated organizationally and administratively to its more established brother service; perhaps more troubling, the latter was suspected by many in the aviation community of an abiding cultural resistance to the airplane. Recall, for example, Colonel Muruzi's report to the Department of the Air Force in which he claimed that there was developing the feeling among the personnel of the Sevastopol School and the aviation squadrons that the commanders of the aviation stations, "not being aviators, will treat aviation with prejudice."[80] The seeming embodiment of these tensions, at least for impatient elements within the popular press who never failed to emphasize this point, was again Gen. Kovan'ko; a longtime aeronaut, he was perched in a seemingly unassailable position of command over the new corps of aviators. Yet it is clear from contemporary and subsequent accounts that he enjoyed the professional respect and admiration of the bulk of aeronauts and aviators alike.[81]

The clash between aeronauts and aviators regarding rank, authority, and expertise was more vividly manifested in the contentious process of selecting the aircraft that would stock the new aviation squadrons. Appraisal of new aircraft systems was conducted primarily by the Aeronautics Committee, a special purchasing and evaluation organ of the Engineering service. It had previously been concerned solely with aeronautical and dirigible technology, experience which again was only partially transferrable to aviation. Reorganized with the transfer of aeronautics to the General Staff, the committee nevertheless largely retained its previous composition; its ranking members continued to be fairly senior members of the engineering services, while it was junior aeronautical and aviation officers who were frequently commandeered to perform consulting and observation missions regarding specific systems.[82] The War Ministry also regularly sponsored competitions which included both domestic and foreign designs in an attempt to evaluate the qualities and capabilities of various aircraft.[83] The staffs of the schools also performed testing and consulting of new inventions and equipment, at times upon request of the committee, which often deferred to their practical expertise.[84] Yet the power of decision ultimately rested with the committees and the higher organs of the army.

This could be, and was often perceived to be, a powerful source of dissatisfaction for pilots of the line. Again, the voices of such pilots are somewhat elusive, but there was no shortage of those who stepped forward claiming to speak for them. Emblematic was the controversy surrounding the Deperdiussin monoplane. A regular feature in *Razvedchik*, "Army Notes," forcefully addressed this issue in late May 1914.[85] Observing that "the state and society donate millions to the conquest of the air, which has made progress with fantastic speed," the piece noted that newspapers recently had reported that the question of military aviation construction, "which excited not only specialists, but broad circles of Russian society" had been resolved. Noting that "it was decided to build a squadron of 326 airplanes of the most diverse systems," the author asked rhetorically if such diversity did not indicate that the question was not at all decided, that "the institution, which is in charge of the fate of Russian aeronautics, is found in a period of searches, doubts, and vacillation." The author was willing to allow, however, that the forthcoming purchasing program was "only an *experiment*, broadly staged. And from this will come good." He had far more fundamental concerns "regarding the *competence* of that board" which decided the question "of state importance" of the aerial building program:

> it is known that all higher places in the aviation units are occupied by persons either completely foreign to aviation or 'aeronauts' (on 'bubbles') persons doubtless knowledgeable in their own field but acquainted with aviation only theoretically. It is not difficult, therefore, for them to be mislead by designers, among whom not all are inspired by the idea of unselfish service to the motherland... for many it's even alien. It is permissible to ask: in the commissions working out the program, are there sufficiently fully represented the *true victors of the air—the military pilots*, who hourly risk their lives in their official capacity and, rising into the celestial heights, answer with their own head for the mistakes of the `earthbound' commissions. Is a *decisive* voice given to them, people of actual experience, people of lofty selflessness and courage?

The author emphasized that "I have the right to ask this question because the first commission, which nearly endowed the Russian army with the poor Deperdiussins, turned out to be not entirely competent. Only by the unanimous protests of persons of practical experience, the press, and society was the unfortunate choice of this system halted." He went on to correctly locate the root of the larger problem: "the whole trouble is that the institute of military pilots is young; the most senior of them, it seems, possess a rank not higher than staff-captain; but by the established military- hierarchical system this low rank is not granted the full right to have its own opinion. However, for such an exceptional matter an exception could be made to the general rule." The author punctuated his plea by appealing for "a wider road to the bold!"

He further indicted another facet of the system whereby pilots were subject to the decisions of those non-aviators above them, namely the two year term of service in the aviation squadrons. Detailing the rigors and demands of aviation service, he noted that for many such officers the reward for their unselfish work would be a transfer back to their regular units after two years. He thus dejectedly added that "I completely understand the pessimism of my 'aerial' friend, who wrote: 'Is this good? The state takes the best powers of its sons, it destroys their health for its own good, it daily exposes their life to mortal danger, in order to then... throw them overboard like unneeded cargo. Posts [of command] in aviation are few, but they are occupied by 'aeronauts,' who place and promote only their own.'" The author noted that if this was so, "it remains only to wish, that our young pilots will age as quickly as possible" and ultimately "will themselves take in their own hands that cause which they, risking their lives, serve. Only then will it break loose from the death loop of dilettantism."[86] He thus concluded with a ringing challenge: "gentlemen aeronauts, step aside a little... Make way for the pilots!"

Again, as with most of the aviation minded pieces which appeared in the pages of *Razvedchik*, this article evidenced its fair share of exaggeration and inaccuracies; it is manifestly clear that the author in question was not an aviator, while it is not at all clear who his "aerial friend" was. Yet his comments provide valuable insight into the climate of opinion, even if not entirely informed, regarding the aviation and aeronautical services, both within these services and within the officer community as a whole. Notably, these comments attested to the real and perceived tensions between the two services, manifested in conflicts between the level of the leadership and that of the line. In like fashion, another contributor to *Razvedchik* who advocated the wholesale development of airplanes at the expense of dirigibles questioned the ability of the responsible military authorities to pursue this path.[87] The author's lack of faith in them was revealed in his account of what surely must have been an apocryphal conversation with a commander "in the air," in which he asked this officer how exactly the dirigible was favorable to the airplane in light of all of the apparent advantages of the latter. The commander allegedly responded that

> 'a dirigible, you see, is more stable than an airplane; flights on the former are notable for their great calmness and certainty; therefore, if it got into the head of the commander in chief to personally conduct aerial reconnaissance, it would be much more comfortable and calm on a dirigible.'
> 'Only this?'
> 'This is all'[88]

Such were the perceptions of the inherent tensions in the relationship between aeronauts and aviators. Realities were often rather more complex. For example, as noted earlier, rather than a thoughtless and callous policy of throwing officers aside like garbage, the two year limit on service in the aviation squadrons was established with a definite purpose. In part it was designed to train a larger number of pilots and thus more quickly build up a reserve pool, but it was adopted primarily because conclusions taken from the French experience held that the health and nerves of a pilot could only withstand the rigors and stress of aviation service for this amount of time. This must have been small consolation, however, to an officer aviator who felt manifestly fit and willing to continue his service in the air forces and yet was not given that choice.

Likewise, the fundamental shortage of posts of command in the air forces was less a function of nepotism among aeronauts than it was the fact that aviation itself was so novel, while aeronautics was already established institutionally. There simply were not that many places or opportunities to promote either aeronaut or aviator to higher command. However, this was still no doubt vexing to devoted and ambitious officer aviators. Again, Gen.

Shishkevich himself, commander of the Aeronautics Unit, recognized how this dearth of opportunities for advancement could negatively influence the morale and service of such officers. It was in part why he proposed to recast fundamentally the present organizational structure and instead create a truly separate aeronautical service arm, complete with its own higher training institutions and posts of higher command.[89] Such reforms, however, were not realized before the outbreak of World War I.

Lastly, regarding the matter of the unfortunate Deperdiussins, it is noteworthy first that the author exhibited the familiar anti-capitalist suspicion regarding those aircraft designers and entrepreneurs to whom patriotism and love of the fatherland was "alien"; it was the existence of such dubious persons that rendered the questionable competence of the selection committee all the more problematic. In fact, however, the Deperdiussins in question had been duly evaluated and selected by the responsible military committee; competing amongst other aircraft systems, the airplane in question performed well, and fit all established War Department criteria.[90] If anything, an examination of the criteria attests not to incompetence but to frequently differing priorities between pilot and command, as well as the rapidity of aviation development itself. The evaluation committees had their own ideas of what an aircraft could and should do, and these did not necessarily include what a pilot himself would prefer if asked. Again, the same process occurred throughout Europe and affected the various air forces in much the same way. In the most notable example, evaluation and purchasing committees, envisaging aviation primarily in terms of reconnaissance and related functions, often sought stability at the expense of speed and maneuverability in aircraft.[91] Such slow, very stable, and very unfortunate aircraft, usually employed quite effectively for artillery spotting, comprised the bulk of the kills for the aces of World War I. At least the Deperdiussins avoided this fate. The author of the *Razvedchik* piece was wrong; the Russian military did indeed go ahead with an order for this system. However, with the outbreak of war, these planes quickly proved too slow, light, and poorly suited for observation; the army halted procurement after 63 planes.[92]

Even being asked did not necessarily guarantee a "correct" choice. The criticism voiced in the *Razvedchik* piece presupposed that there was indeed universal agreement among pilots of the line about such issues. Yet we have already seen that this was far from the case, particularly in the contentious welter of disagreement and intrigue that surrounded the choice of specially modified combat Nieuports for the Russian aviation squadrons in 1912. In many ways the situation was the exact opposite of the one regarding the Deperdiussins. Instead of a committee of out of touch aeronauts and aging engineers, these Nieuports had been personally and dramatically tested by two renowned military aviators, Andreadi and Dybovskii. It was their heroic example which subsequently impressed and convinced the Grand Duke Alexander Mikhailovich, whose influence was enough to see to it that the Nieuports were adopted by the army. Yet, as Col. Odintsov had complained, numerous parties harbored serious reservations about the Nieuports, including two of his then subordinates, Capt. Prince Muruzi and senior instructor Staff-Capt. Viktor-Berchenko.[93] Such reservations were prescient; by 1916 it was admitted that the choice of the Nieuports had been, in the words of Col. Nemchenko, a "terrible mistake."[94]

Perhaps the quintessential example of these multi-faceted professional cleavages between aeronauts and aviators, between those in command and those in the field, was provided yet again by Captain Petr Nesterov. Some of his exploits, such as his flight from Kiev to Gatchina and his early experiments with field units to use aircraft for artillery spotting, have already been noted above. He is most famous, however, for being the first aviator in the world to perform an aerial loop, two days before the Frenchman Pegoud, in Kiev in August 1913. Nesterov's loop was exactly the type of radical maneuver considered reckless and impossible and thereby eschewed by the early aviators and many of his military comrades.[95] Yet Nesterov had been trained to fly several years later, and by then both the theory and technology of aviation had made significant advancements. His feat was nevertheless revolutionary. The newspaper *Kievlianin* hailed Nesterov's achievement in rapturous terms as a breakthrough for the world of aviation, completed in Russia by a Russian on a plane built in Russia.[96] In contrast, Capt. Nesterov was temporarily "placed under arrest for 'undue risk with a machine, the property of his government,'" as he had not received official permission to attempt such a feat.[97] In an interview regarding the incident Gen. Shishkevich is alleged to have commented that "experiments by military pilots like the one done by Nesterov are useless. These acrobatic tricks find their justification in the performance of professionals, but I am surprised how the command permitted this officer" to attempt this.[98] Yet within a few days Nesterov's arrest was vacated, he was promoted to Staff-Captain, and in November was awarded the Imperial All-Russian Aeroclub Gold Medal.[99] Nesterov himself, in a letter to the French newspaper *Matin*—itself illustrative of the cosmopolitan world of aviation—explained the motives and goals regarding his death loop. He noted that he had begun considering the loop long ago and had thoroughly developed his theories, while radically different from commonly accepted ones, through observation of birds and hypothetical formulations. He concluded that "my experiment was undertaken only when I was completely sure of its success. I did not want to and could not risk my life, as I am the father of two children, a young daughter and a young son, future aviators."[100]

As with questions of policy, institutions, and command, the processes of recruitment, training, and service in the aviation services exacerbated familiar tensions and divisions, threatened to destabilize vested constituencies and legitimate new ones, and gave urgency to questions of competence, professionalism, and identity. At stake was a unique paradigm and ethos of service for the new community of *voennye letchiki*; it is not too dramatic to say that this constituted yet another front in the war for the soul of the aviation services. This was a war fought on various fronts by numerous parties. Policies regarding recruitment and personnel revolved around the search for the ideal officer-aviator. But there were competing perceptions of this ideal, at the levels of institutions, command, and the line aviators themselves. Notably, judging from memoirs and the early lists of graduates, it appears that the Grand Duke and the Department of the Air Force did indeed pay some heed to issues of prestige of unit and aristocratic pedigree in admissions to the Sevastopol Aviation School.[101] Yet in their decision to accept volunteers from all branches of arms, it seems the Grand Duke and his colleagues ideally sought officers who would combine the best of education and training with proper breeding, drawn from an all-Empire constituency. Reminiscent to an extent of the successful model of modernization within the German officer corps that appealed to many Russian officers, and true to the proud historical tradition of the Russian officer corps itself, it was an ideal of recruitment which not surprisingly fit the social and imperial predilections of a Grand Duke.[102] Even this predilection, however, is hard to detect in the graduation lists of the Sevastopol Aviation School from 1913. With the standardization of admissions and the rapid growth of the air forces, the Sevastopol School increasingly attracted a more varied constituency, largely reflecting the increasing attraction and significance of military aviation itself. Allegations of exclusivity and social bias in admissions would persist in the press and particularly in memoirs, but the evolution of aviation was ultimately a larger and more powerful phenomenon than the defense of social and status prerogatives.

The same process was at work in the Officers' Aeronautics School and its Aviation Section. Whether one believed the numerous charges of outsiders that the School had never truly reconciled itself to a new and independent culture of aviation, or whether one believed the School's own protestations of good faith and commitment to the airplane, the rapid expansion of aviation created its own imperatives. Thus, the School inevitably underwent its evolution from the preserve of engineers to a more diverse composition as officers from all branches of arms flocked to the Aviation Section. The staff of the Aviation Section itself increasingly reflected not its aeronautical parentage, but a cadre of officers whose training and commitment was devoted exclusively to aviation.

This larger expansion and evolution of the aviation service fed into the process of training. Regarding policies of recruitment, failure to volunteer, refusal to accept a posting, and complaints of exclusiveness and obstruction were largely reactive responses by officers to policies devolved from above. In contrast, in the realm of elaborating a regime of training, the staffs of the schools and the officer volunteers themselves played a far more active role. Yet this did not ensure conformity or harmony. Conceptions of the proper training, role, and function of a military aviator ranged the spectrum from more caste-informed notions of fraternity and innate talent to fundamentally professional models based upon rigorous instruction, education, and technical expertise. These competing ideals revolved to a great extent around the institutional and cultural divisions between the Sevastopol Aviation School, imbued with a sense of mission and dedicated to rapidly providing the Empire with both the instruments and cadres of an air force, and the Aviation Section of the OVS, arguably devoted to a broader and deeper understanding of what it meant to be a military aviator. Moreover, individual officers at times resisted, at times evaded, the prevailing ethos that these aviation schools sought to inculcate. Many of these volunteers brought to the schools their own compelling motivations and conceptions of what it meant to serve as a *voennyi letchik*, whether it be the search for glory or mere escape from a regimental chancellery. Such conceptions did not always match those of their instructors and commanders. Again, the explosive expansion and evolution of military aviation, the rapid term of pilot training, and the very youth of the air forces ensured that socialization and acculturation to any paradigm of aviation service was an uneven process within the setting of the schools.

Upon graduation and transfer to line aviation units for the *voennye letchiki*, practicing their novel craft and asserting their vital role and status entailed navigating prevailing norms of hierarchy, deference, and duty specific to the community and culture of the Russian officer corps as a whole. The process of establishing a novel profession of arms was thus a contentious one, frequently pitting pilots against bureaucrats, aeronauts, the General Staff, and other strangers to aviation. The complaints voiced in the military press, the dramatic gestures by aviators, the clues they left through orders and reports, all implied that only identity of training and service could provide identity of interest for the cadre of *voennyi letchiki*, and that only these military aviators themselves could protect and further those interests. Even then, in light of the rifts—educational, institutional, generational—within the community of military aviators, this was no guarantee of unanimity of thought regarding what it meant to serve as a military pilot. The exercise of initiative, the assertion of special interests, and the questioning of authority could and did produce tensions within the ranks of aviators themselves, between professionals and poseurs, between serious aviators and fashionable dilettantes. However, all of the fissures, cleavages, and conflicts which pitted aviators both against themselves and against other vested interests and constituencies were couched in a common language. Regardless of one's position in these numerous debates, the war over the future of the Russian air forces was phrased in terms of a struggle of cultures, between cultures of voluntarism and inertia, initiative and established routine, competence and deference, selfless service and the narrow furtherance of parochial interests.

The vagaries of recruitment, training, and service thus produced disparate possibilities for the elaboration of an ethos of service for the *voennye letchiki*, but the Russian air services were most characterized by an emerging and urgent professionalism. Dilettantes, poseurs, and *beloruchki* there were, but theirs was largely a negative example. They appeared in reports, memorandums, the contemporary press, and subsequently memoirs as objects of derision, not as standards of an alternative and valid paradigm of service. While many such officers clung tenaciously to their own caste-informed notions of role and duty, and while the perceived hindrances they represented to professionals were at times vexing, they were not vocal advocates of a competing ideal, but passive resisters to a dominant one.

In contrast, the predominance and persuasiveness of the professional ethic were again exemplified by Mikhail Efimov and Staff-Capt. Petr Nesterov. They advocated a thoroughly professional ideal of service, emphasizing sound theoretical and technical education, rigorous standards of admission and training, a dedication to constant inquiry and self-improvement, a palpable sense of self-esteem and group identity, and a marked recognition of the special interests and demands of aviation. They preached their gospel in the pages of the military, aeronautical, and popular press, gave power to their words with acts of daring and innovation, and were cited in orders that cajoled and exhorted their fellow aviators to follow their example. Their voices were augmented by like-minded colleagues and fellow proselytizers who appeared in progressive journals such as *Razvedchik* and *Vozdukhoplavatel'*, citing the example of Efimov and Nesterov in calling for a uniform and empowering professional ethos for officers who devoted their careers to aviation.

At the same time, the professional ideals of Nesterov and Efimov attested to the broader evolution of aviation. While they trained and served at two markedly different schools, with quite distinct concepts of mission, the professional ties that bound the two men transcended these and other divisions of officer and civilian, education and background. Despite the social predilections that prevailed at the Sevastopol School, Efimov absolutely insisted upon the importance of theoretical knowledge and technical expertise in the training of aviators. In turn, Nesterov, who due to his theoretical studies and experiments in aeronautics felt far more at home among the like-minded staff of the Officers' Aeronautics School, also focused his prodigious energies and talents upon the practical development and application of the potential of the airplane. Both men represented, as did more and more of the instructors and trainees of the two schools, a new, more democratic culture of aviation, one expressly devoted to the demands and imperatives of the airplane. Thus, upon Nesterov's visit to the Sevastopol Aviation School in March 1913, this "ardent admirer" of the first Russian pilot finally had the chance to meet his illustrious colleague. Efimov congratulated Nesterov on his loop and its great significance for aviation, while the two men then actively questioned each other on such matters as pilot training methods and flying in strong wind.[103]

The specific demands of recruiting, training, and posting a cadre of military aviators could inspire far broader demands, challenges, and critiques of prevailing notions of duty, service, hierarchy, and discipline in the tsarist army. Yet there was far more to creating a community of *voennye letchiki* than the requisites of learning to fly. In their aircraft and in the sky, these aviators remained military servitors, a complex and multi-faceted identity. These aviators would thus exert a profound influence upon the cultivation of a culture of Russian military aviation not merely as pilots, but as tsarist officers.

7| Officers, Gentlemen, Citizens

The challenges involved in defining and inculcating a standardized regimen of aviation service were varied. Even narrow technical demands and requirements could spawn broader controversies regarding standards of expertise, authority, and discipline, all distilled within the milieu of competing visions, institutions, and constituencies explored in previous chapters. Yet problematic as this process was, defining the role and duties of a military aviator involved larger issues than the ostensible requirements of effectively flying an airplane. The *voennye letchiki* were officers as well as pilots, and as such their experiences of recruitment, training, and service were influenced by broader concerns endemic to the tsarist officer community as a whole. Such concerns were often magnified in the context of abiding anxieties over the composition, quality, and loyalty of the officer corps' cadres following the experiences of 1904–06.

Thus, official policies of recruitment, training, and service were in part informed by such considerations as nationality and branch of service, while issues of social background, patronage, and honor were often determining aspects of the environment of the aviation schools and squadrons. At times the assumptions which underpinned official policies resonated with the diverse constituency of military pilots; at other times officers adhered to their own codes of conduct to subvert both the letter and spirit of regulations devolved from above. These broader concerns served as sources of both unity and dissonance amongst the corps of officer aviators, and were instrumental in the elaboration of a specific culture of Russian military aviation.

Duty and Loyalty

A variety of factors beyond the immediate confines of the cockpit influenced the pilot training regimens of the tsarist aviation schools and the collective identity they sought to impart to their trainees. Notably, pilots were officers, but there was nothing inevitable about the equating of the two. Mastering the art of flying did not inherently require a certain rank in the service lists or epaulets on one's shoulder. As such, the officer-pilots of the line were faced with the task of distinguishing themselves from those with potentially competing claims to their special status.

The immediate if seemingly unlikely source of such claims was common soldiers. This phenomenon was a product of the army leadership's early decision to train pilots from the ranks and include them as part of the regular complement of an aviation squadron, which was to contain four officer-pilots and two soldiers. The primary motivation for this decision was again numbers and need. Given the rapid expansion of aviation squadrons foreseen for the near future, large numbers of trained pilots would be needed. Yet in light of continuing problems with volunteering and recruitment, it was recognized that the officer corps itself was unlikely to entirely fulfill that need. The War Department order that permanently established the Aviation Section of the Officers' Aeronautics School thus foresaw training 10 soldiers to fly in each class.[1] It is significant, however, that despite being slated to receive the same training and instruction and being required to pass the same examinations as officer-pilots, pilots from the ranks were not to be granted the designation *voennyi letchik*, while their material incentives and remuneration were of course less substantial as well.[2]

Nevertheless, the Russian military leadership made a conscious and telling decision not to retain aviation as the exclusive preserve of officers or "gentlemen." In light of the debates concerning the nature of flying and piloting, the authorities clearly came down upon the side of those who considered aviation a skill to be learned and mastered. At least in theory, those with intelligence and enthusiasm could thus qualify, regardless of their place in the military hierarchy. The French and German armies also envisioned an air force which would depend in part upon soldier pilots, but the Russian military leadership anticipated perhaps a more widespread use of them.[3] In reality, however, the training of pilots from the ranks, although provided for in the plans and regulations, was rather spotty in the Russian aviation services. A report of the Main Engineering Administration to the War Minister in March 1912 spelled out in detail the concerns regarding the training of aviators and especially the rationale and results of training soldiers as pilots.[4] The report noted that according to the regulations of the Aviation Section of the OVS it was planned to assign yearly 15 officers and 20 soldiers from the ranks to train there, of which 10 of the latter were to be trained as pilots and 10 in aircraft repair and maintenance.

However, the report noted that in light of the planned acceleration of the formation of aviation squadrons to 53, instead of the original 18,[5] there was a large shortage of pilots and specialists from the ranks.[6] Moreover, in order to simply replace the normal losses of pilots, it was necessary to yearly send to the schools for training not less than half of the total number of all pilots presently assigned to aviation squadrons. It was thus necessary to train 106 officers and 53 pilots from the ranks per year, "as by the experience of France their pilots can be in aviation service for only about two years." Even worse, "the experience of France has shown that of 8 [persons] training on flights in the school, barely 5 complete training on Farman aircraft, and on monoplanes, especially the Nieuport aircraft, the number of those who have finished training will be even less."[7] The report continued that the War Minister had thus already ordered the training of officers to be expanded. In contrast, "as regards pilots from the ranks, those up to the present time we do not have, unless 2-3 random motorists from the ranks trained to fly at the Aviation School of the OVF are counted."[8] In light of these considerations the Engineering Administration considered it necessary in the present year to assign to the Aviation Section, instead of the 25 soldiers established by regulation, two for each of the 35 officers training, namely 70 soldiers, from the most able for aviation service, by selection of the local command; 30 would be trained to fly on planes, the rest to service aircraft. The report proposed to obligate the Department of the Air Force to do the same, to train 30 soldiers to fly yearly and the rest to serve in maintenance and repair. It concluded that in addition the Engineering Administration felt it necessary to submit to the War Council a proposal regarding the organization of new aviation schools to train 106 officers and 53 soldiers to fly yearly. The report did note, however, that "schools for the training of the lower ranks, in view of discipline, perhaps will seem more convenient to organize separately from the officers' schools."[9]

Although records are far less systematic and clear than for officer-pilots, available evidence indicates that the concerns voiced in this report were not assuaged. A report of the Officers' Aeronautics School to the Chief of the General Staff, dated Jan 16, 1912, noted that it was forwarding a list of 28 officers finishing the Aviation Section in 1912; it recommended that all receive the designation of *voennyi letchik*.[10] In contrast, following this report in the files is a list of 46 soldiers assigned to the Aviation Section as of March 1913; of these 46, three were being trained as pilots, the rest as joiners, metalworkers, motorists, and mechanics.[11] In a report of May 1913 to Chief of the General Staff Zhilinskii, Gen. Shishkevich noted that there were then only 10 trained soldier pilots in the 18 existing aviation squadrons, while these squadrons possessed their full complement of officer pilots.[12]

Evidence from the Sevastopol Aviation School further suggests that the training of pilots from the ranks faced passive yet effective resistance on the part of officers. Even a successful example of such training indicates that it depended to a large extent upon the willingness of individual officers to teach the personal mechanic assigned to their plane. This is indeed how most soldier-pilots learned to fly.[13] A short piece in *Sevastopol Illustrated Aviation*

Journal in fall 1911 entitled "First Pilot from the Ranks" noted that "with great pleasure we mark the fact of the appearance in the Russian army of the soldier-pilot."[14] He was Dmitrii Semishkura, of the Preobrazhenskii Guards Regiment, who on October 28th had passed the examination for pilot. Private Semishkura had come to the school along with Staff-Capt. Viktor-Berchenko; "according to the latter he became interested in the matter and, constantly found at the plane, learned the apparatus and motor well." Thus,

> in reward for his zeal, ardent attitude to the matter and vigilant care of the airplane, supervisor Staff-Captain Viktor-Berchenko, a little at a time, at odd moments, began lessons with Semishkura. These lessons led to an excellent result. The Preobrazhenskii Life-Guards Regiment, besides a pilot supervisor in the person of Staff-Captain Viktor- Berchenko, has gained pilot private Dmitrii Semishkura.[15]

Yet not all officers were as well-disposed to their mechanics as Staff-Capt. Viktor-Berchenko. The paltry numbers of trained soldier pilots indicates that apparently there were numerous officers who were in no rush or mood to recognize the role of pilot as the proper domain for a non-officer, much less to help train them for such a function. It is clear that this resistance did not solely revolve around concerns of competence, education, or qualifications. The Engineering Administration report cited above itself suggests why the training of soldier pilots was lagging despite official policy: the admission that 'in view of discipline" it might be better to separate officers and soldiers was revealing. An analogous process was apparently at work in the British air services; despite plans for a mixed force of officers and soldiers, the absence of firm policy decisions produced a cadre of pilots who were almost all commissioned officers.[16]

The disdain of many Russian officers for the notion of pilots from the ranks inevitably raises the issue of class divisions. As noted, various scholars have asserted that the Imperial army in many ways merely replicated in uniform the caste relations of noble and peasant that prevailed in tsarist society and culture as a whole.[17] Regardless of statistical realities, the alleged aristocratic orientation of the Sevastopol Aviation School was a particularly contentious example of the problematic relationship of social background and professional identity. The perceived contrast between the social composition of the two military aviation schools became inextricably bound up in the debate between pilots as pure fliers or technically proficient aviators. Thus, despite the presence of Efimov, a student of aviation par excellence, it was allegedly dilettantes and poseurs who regarded aviation as an attractive fad that were attracted to the Sevastopol School and its emphasis upon flying, while the more industrious and serious-minded pursued their careers through the Aviation Section of the OVS and its theoretical and technical orientation. In the former group were primarily stylish scions of the nobility and upper classes; in memoirs and popular histories, the term *beloruchki* is usually preceded by *dvoriane*.[18] While allegations and the truth are not necessarily harmonious, a marked dichotomy between the devotee and dilettante was a phenomenon common to the the new military air arms throughout Europe.

These divisions spilled over into the relations between officer-pilots and their mechanics, mostly young NCO's and privates, which meant men largely of the lower classes. Worse than the passive resistance of not teaching soldiers to fly, the aristocratic bent of the Sevastopol School allegedly manifested itself in outright class antipathies between officers and soldiers. Regarding this issue, Soviet histories and memoirs, given the ideological imperatives at work, must be treated with the appropriate grain of salt, but that does not mean they are without valuable insights. Such works invariably make a clear distinction between two types of officers. They speak fondly of those "progressive" officers, including a few with aristocratic origins, who treated their mechanics with respect, who discussed technical matters with them and actually helped service the aircraft, and who at times even taught their mechanics to fly. As with Viktor-Berchenko and Semishkura, so it was with the memoirist Ivan Spatarel' and several of his comrades. He was trained to fly by the officer he served, Staff-Capt. Zemitan, who himself trained under Mikhail Efimov; not surprisingly, Zemitan displayed an avid interest in technical matters, and was subsequently appointed to the staff of the school.[19] At the other extreme were those officers, some of them instructors, who would merely appear before their plane, barking orders frequently laced with curses, and who would not be caught dead with their mechanics, much less fixing a plane.[20] For such officers teaching their mechanics to fly was out of the question, nor did they look kindly upon those officers who did so. Spatarel' notes the outrage and scornful looks Zemitan provoked amongst his fellow officers while working with him on the aircraft and motors. Not surprisingly, the professional interests and flying abilities of such officers are usually disparaged in the literature. Yet it is worth noting that despite its allegedly more progressive outlook, the OVS also largely failed to produce soldier pilots.

Moreover, there were those officers who zealously guarded their social prerogatives and yet were conscientious and thoroughly professional pilots. One could be a devoted, diligent aviator and an arrogant, haughty officer at the same time, as was apparently the case with Staff-Capt. Rudnev, the first Russian military aviator and longtime instructor at the Aviation Section of the OVS.[21] The name which most frequently arises in the memoirs in this regard, however, is that of Viacheslav Tkachev, a Cossack officer at the time of his appointment to the Sevastopol School and a future commander of both the Imperial air forces in World War I and the White air forces in the Russian Civil War. As Spatarel' describes him, Tkachev possessed incredible skill and powerful presence as an aviator, as well as thorough technical knowledge and training; nor was he afraid to dirty his hands, and he had no patience for those officers who disdained technical matters or their mechanics. In his relations with officers, especially with aristocratic *beloruchki*, Tkachev was "independent, coldly polite," while with soldiers he simply paid no attention to them, "as if they didn't exist." Yet Spatarel' notes that in this they sensed power, and that they both respected and were somewhat afraid of him.[22] In the same vein, it is worth noting that Spatarel' later recounts that during the war Tkachev was far more willing than other front air commanders to promote outstanding soldier pilots to the rank of officer, not because he had any great love for them, but because he duly recognized and rewarded professional ability and accomplishment.[23] Other sources corroborate that throughout his career Tkachev enjoyed a reputation as a strict yet just officer and commander, respected by his subordinates and superiors alike.[24]

All of this attests to the fluid nature of the relationship between officer pilots and soldier mechanics. Providing this relationship with form and content was to an extent a matter of personal preference and predilection for an officer. Some of these relationships were thus defined and cemented around professional demands and needs specific to aviation, while others derived from the dynamics of the master-serf attitudes that found expression in the officer-soldier relationship in the army.[25] Some officers emphasized their social status and prerogatives, others made common professional cause with their valuable assistants. Considering how important a mechanic was to not only the successful work, but thereby the continued existence, of a pilot, the latter course was probably the best option.

Regardless of the stance taken by an officer toward his mechanic, the title of *voennyi letchik* did not necessarily obliterate the identities and affiliations that officer pilots brought to their new branch of arms. If the identity of military aviator was a powerful one, it was frequently augmented by other affiliations—of branch of service, of unit and regiment—which continued to make themselves felt. Some officers attempted to seamlessly reconcile their new role with their old. In a petition to the commander of the 1st Life-Guards Artillery Brigade, Lieut. E. A. Nesterov noted that having finished aviation school "and wishing to continue my service in aviation," he was requesting to be posted to the aviation squadron of the Guards Corps. He explained that "since all of my service in the ranks of officers has been done exclusively as a member of the Guards Corps, it would be extremely desirable for me to continue it there, serving the cause of aerial reconnaissance and not losing connection with the unit in which I have served from the day of my promotion."[26] Yet previous service experiences and affiliations could also prove detrimental to efforts to cultivate an ethos unique to the air forces. In his larger plan for the recruitment of military aviators, Col. Odintsov had indeed pointed to the inadequacies of the current system of culling volunteers from the regiments, and instead had emphasized the need for a corps of flyers "from a young age trained and raised in the spirit of their future service." Although his rationale specified the inability of ordinary officers to absorb the fine points of flying after several years of service, what he implied was something larger than simply mastering the techniques of piloting. Instead, Odintsov seemed to be emphasizing a failure to fully make the transition to a style of service and duty that was very new, different, and unique. Hence his proposal to form a military aeronautics secondary school [*uchilishche*] that would accept students from both the cadet corps and civilian *gymnasia*.

There were larger defining forces at work in the Russian Imperial officer corps, not all of which could be manipulated at the level of personal choice. Issues of ethnicity, nationality, and religion also became inextricably entangled in the process of cultivating a cadre of officer-aviators. Of particular salience was the Russian army's project to develop quotas for nationalities in the composition of the officer corps following the events of 1905. Drawing in part upon precedent and existing rules, the project also sought to expand restrictions upon various nationalities in regard to service in technical units and sensitive branches of the army. Such restrictions were to be extended also to the new aviation service. While this project was in committee and the existing restrictions under review for most of the period before World War I, its tortured evolution eventually caught up with the cadre of officer pilots.[27]

In secret correspondence, dated April 27, 1913, from the Aeronautics Unit to Gen. Kovan'ko, it was noted that in the training class of the OVS there were three officers—Lieutenants Shomanskii and Nagurskii and Second Lieut. Bashko—who were of Roman-Catholic faith and upon whom the projected regulations forbidding non-believers [*inoverstev*] from aviation service should be applied. These officers were thus subject to dismissal back to their regular units. The memorandum noted, however, that "owing to the strong desire expressed by these officers to serve in the aviation squadrons" the Chief of the General Staff had permitted them to remain at the school, on the condition that Shomanskii and Nagurskii convert to Orthodoxy (at their suggestion) and that Bashko agree to serve in Siberia, Turkestan, or the Caucasus. The Aeronautics Unit thus asked Gen. Kovan'ko to inform them of the necessary dispositions and to submit Bashko's written pledge.[28] The requisite report submitted by

Lieutenant Shomanskii of the Siberian Rifle Regiment to the commander of the Aviation Section was particularly revealing. Although a petition composed with a definite purpose, it was a notably poignant plea, touching upon issues of both individual and group identity. It is worth reproducing in full:

> my father and mother are descended from nobility of Volynsk province; when I turned three years old our whole family moved to Ekaterinoslav province, where it still lives now. When school age came, my parents, fearing that the Polish language could hinder my studies in Russian, impeded me from studying and even speaking Polish, as a result of which I understand Polish very poorly and almost cannot speak it. There was no permanent Roman Catholic Church or *ksendza* in the city and only once a year did the clergy come to perform religious ceremonies, owing to which there did not develop in me an attachment to the Roman Catholic Church. Upon entry into a military *uchilishche*, I fell into an exclusively Russian environment, and having become close friends with my comrades, I gradually grew accustomed to Russian Orthodox rites and more and more isolated myself from Roman Catholic doctrine. With promotion to officer I joined a regiment which became for me a second family. The influence of a friendly, comradely environment and the whole regimental life united me completely to the Orthodox religion and in my heart I stopped being Catholic, as a result of which at the present time I request your intercession for permission for me to switch from the Roman Catholic religion to the bosom of the Orthodox Church.[29]

Second Lieut. Bashko also duly provided a written pledge to convert.[30] Lieut. Nagurskii's petition to serve without restrictions—which he justified "by the fact that I was enrolled before the publication of these restrictions and that I have already been flying for about a year"—having been denied, in a report to the Commander of the Aviation Section he petitioned to remain at the school on condition that he would agree to serve in Siberia, Turkestan, or the Caucasus.[31] In a subsequent report to the Aeronautics Unit Gen. Kovan'ko noted that Nagurskii had received permission to stay on the condition that he convert, but "apparently Lieut. Nagurskii has once again changed his wish," and thus he was forwarding Nagurskii's report for further consideration by the Chief of the General Staff.[32] The newly proposed rules affected other officers who had completed training and were already serving in aviation squadrons. In a telegram to the Aeronautics Unit a Capt. Nizhevskii, serving in Berdichev aeronautical company, complained that he had heard his transfer to Bronitsa was denied due to religion. He explained that "although I am a Catholic I am a native of Tambov province and married to an Orthodox" and thus asked for help to have him transferred to Bronitsa or another aviation company in European Russia, for example in Warsaw, which he had heard was free.[33]

An even more serious affair cropped up in late 1913 when the Aeronautics Unit, in light of the new aviation squadrons being formed, sent out inquiries to the aviation schools and units regarding "officer-pilots who, by their moral and service qualities," merited promotion to squadron commander. However, as the Aeronautics Unit subsequently informed the Chief of the General Staff, when the requisite information was collected it turned out that among these officers there were three Catholics, Staff-Captains Gornovskii, Zaleskii, and Zagurskii, whose appointment as squadron commander had been solicited by the commander of the 1st army corps, General Artamanov. There was a *karait* as well, Staff-Capt. Saatchi-Samadchi. These officers had all been trained at the Sevastopol Aviation School and graduated during 1913. The report noted that these officers had entered the school before the restrictions regarding non-believers were issued; it thus concluded by asking whether they could be appointed as commanders and if so with any geographical restrictions.[34] From the records, unfortunately, the outcome of this affair is not entirely clear.

Issues of social composition and nationality, particularly when commingled with institutional and occupational affiliations specific to the realm of aviation, could be divisive elements in the process of forging a community of *voennye letchiki*. Yet despite such cleavages, there were threads of unity at work as well within the fledgling aviation community. These unifying threads also derived from pilots' broader identity as officers and the notions of duty and service that status entailed.

Notably, common concerns of honor continued to exert a powerful influence in shaping the perceptions and careers of the officer pilots. As with their brethren in other branches of the army, negative interactions with the public often served to cast such concerns in bold relief. When Gen. Kovan'ko submitted a report regarding his charges of slander on the part of the newspaper *Vechernee vremia*, he noted that a special assembly of the officers of the Officers' Aeronautics School had been called to discuss the incident; he thus attached an excerpt from this assembly's resolutions. An examination of this document is instructive.[35] This general assembly was held under the chairmanship of Lieut. Colonel Prince Baratov, with 25 officers of the permanent and training staff present. The excerpt noted that having discussed the article in question, the society of officers "unanimously" decided upon a resolution which contained eight points. With the first three points, the officers resolved to recognize the article as "an outrageous attack against the school"; to react to this article as an attack against the entire school; and to ask the appropriate establishment to force *Vechernee vremia* to publish a refutation to the indicated article. Point four noted that a committee of three officers—Captain Vegener, Lieut. Kanishchev, and Staff-Captain Sol'skii—were chosen to work out the text of this refutation, which was subsequently approved by the society of officers. It read: "the society of officers of the

Officers' Aeronautics School, outraged by the article... considers it necessary to bring to general attention, that that which is published in it is tendentious slander directed against the school and its personnel." Point five, "recognizing the act of the author of this article, as well as the person who provided information for this article, as disgracing the honor of the uniform," resolved to take measures to ascertain the author or the person who had given information to the correspondent for *Vechernee vremia*. Point six further obligated each of those present to take all measures to fulfill point five and to report anything they might find out to the general assembly. Point seven required all officers, doctors, and officials of the permanent staff of the school and those assigned to it to take a written pledge that they had not participated in the drafting of the article or in providing information for it. Lastly, point eight stipulated that "in order to take measures so that in the future similar articles are not repeated in the press, to declare that an officer who has given information for similar slanders or who has taken part in their drafting disgraces the rank [*zvanie*] of officer."

A less contentious but equally powerful marker of identity and honor for the fledgling corps of Russian military aviators was the distinction between themselves and the small but growing group of civilian pilots, notably those of Western Europe. Here again anti-capitalist sentiments once again provided a clear measure of delineation. A short piece in *Razvedchik* succinctly summarized this distinction when it noted that the best pilots in the world were the French, who having achieved virtuosity were "occupied mainly with records," a pursuit which frequently attained a sort of circus-like character. It noted that "the success of records, accompanied by the clamor of publicity, the ovations of crowds, and monetary prizes," acted like a drug upon such pilots, "forcing them to concoct the most baffling tricks and to forget the practical side of aviation." In contrast, for the military pilot, "his work in appearance is more modest and every success is purely practical in nature." The piece emphasized that "our military pilots, by their long flights, their refinements in photography, their bomb throwing, and their constant practice in reconnaissance work, have shown that their task is understood by them."[36]

Yet such sentiment could be equally exclusionary even within the Empire's borders. Despite their respect for the efforts of exceptional civilian figures such as Popov and especially Efimov to popularize and advance Imperial aviation, and regardless of a certain awareness of belonging to a special fraternity of the air, the officer pilots of the Russian Imperial army and navy were acutely conscious that they were *military aviators*. They carried this official designation with pride, and again the distinction broke down to the juxtaposition of duty and self-interest, selflessness and material advantage.

Anti-capitalist sentiment, shorn of its xenophobic tendencies, thus functioned in another vein as a powerful marker of professional identity and honor for aviation officers. Recall the public funeral of Capt. Matsievich, whose military status exaggerated the tropes of selflessness and patriotic service attributed to his sacrifice.[37] While Russian civilian pilots escaped much of the severity of the criticism directed at their foreign comrades, especially the French, their function was still to perform, to entertain, and to compete for money and other prizes. Such was not the work of officers.

The process of elaborating and inculcating an ideal of service for the *voennye letchiki* involved far more than the particularities of piloting itself. A variety of broader factors came into play, while the officer-pilots themselves were both active agents and passive subjects in negotiating concerns of honor, nationality, social class, and branch of arms. Moreover, these officers would further face the challenge of legitimating their unique role and identity within the broader context of an officer corps struggling to define itself as a whole as it grappled with issues of competence, loyalty, honor, and tradition.

Community and Example

Framed again around issues of military competence, expertise, and authority, the tensions between commanding engineers and aeronauts and the aviators of the line were frequently replicated on a broader level, that of the army as a whole. It was this audience, particularly the levels of senior command, to which aviation officers had to appeal as they sought to establish the credibility and relevance of themselves and their new branch of arms.

Regarding the reception accorded them by their comrades, inertia, apathy, and skepticism As noted, members of this audience could be extremely receptive. To many the onset of aviation could be seen as an extraordinary portent, a promise of power and redemption. A piece in *Razvedchik* in 1911 described in rapturous terms the excursion of the Riga Society of the Zealots of Military Knowledge to the Sevastopol Aviation School. This included a tour of the aerodrome, compete with demonstration flights and later the chance to ride as passengers with the officer-pilots, which left all incredibly impressed with the power of the airplane and the skill of the aviators.[38] In his October 1913 report regarding his flight from Kiev to Odessa and Ekaterinodar, Sub-Esaul Tkachev of the 2nd corps aviation squadron noted that at the request of the ataman and chief of staff of the Cossack host in Ekaterinodar, for three days he performed demonstration flights for the troops and educational institutions and gave a lecture to the officers. He also lectured and answered questions about aviation from students of the local *real* school.[39]

A more influential constituency was the senior line officers, among them the commanders of the military districts and army corps. They too frequently responded quite favorably to the new heroes in their midst. In a secret report to the GUGSh in Oct. 1913 regarding the participation of the 3rd aviation and 10th aeronautical companies in summer maneuvers, the Chief of Staff of the Kiev military district gave a quite negative review of dirigibles while noting that in the majority of cases the airplanes fulfilled their assigned tasks, conducting flights and providing information to the staffs of the army corps.[40] The chief of staff of the 10th army corps was even more enthusiastic in his Nov. 1913 report to Staff-Capt. Nesterov, commander of the 2nd corps aviation squadron, regarding the squadron's participation in the corps' maneuvers. He noted that the aircraft performed "very successfully" and that "Staff-Capt. Nesterov worked marvelously, and next Lieut. Tkachev provided more exact information."[41]

In December 1913 the regular "Army Notes" feature in *Razvedchik* noted that it was already the third year in which aeronautical apparatuses had been used as a new means of strategic reconnaissance at the large maneuvers. It summed up the prevailing feeling on the matter by observing that

> the opinions of senior commanders are diametrically opposed: one side, for example, treated the activity of its aviation squadrons with complete condemnation and denial of their benefits, the other side- with great praise and recognition. This contradiction clearly indicates that the technology of military aeronautics already now stands firmly, up to the mark, and that the entire question consequently lies in personnel and the ability of staffs to use this new 'branch of arms.'[42]

The piece reported far fewer doubts about the *voennye letchiki* themselves: "in general selfless, always involving mortal danger, the work of the majority of our pilots during all the maneuvers seen is above any praise." The article concluded that "in any case, the new 'branch of arms' received full recognition in the line, which was expressed by the huge interest in the flights and the loving attitude toward the pilots. Daring and courage always find fitting recognition." This piece succinctly reiterated themes that lurked in the official reports and appraisals of the aviators and their work. It recognized military aviation as a newly established and potentially transformative branch of arms, provided—a significant caveat—that the army staffs learned to use it. Moreover, a notable distinction was made between the higher levels of command, where opinion was still divided regarding aviation, and the sphere of the line, where the airplane and its pilots had received "full recognition."

For the *voennye letchiki*, however, the process of securing and cementing such recognition among the higher reaches of the army was further complicated by what might be termed the dilettante phenomenon. Aside from those pilots who pursued aviation as a fad, another potential source of exasperation was the activities of those who could not even claim true membership in the aviation and aeronautical services. There were numerous well-meaning, energetic officers who took it upon themselves not only to support but to get actively involved with aviation; while some were powerful or well-positioned enough to exercise influence, they did so without any requisite knowledge or training in this area. As many of these officers were unfortuitously placed in positions of command and influence, their voices could not be ignored, nor could one entirely refuse to curry their favor.

Again, particularly in the very early years of aviation, such initiative often was a positive phenomenon, when local endeavors preceded and/or laid the basis for future official War Ministry efforts. Of course, the best example of a concerned, influential, and amateur patron was of course the Grand Duke Alexander Mikhailovich himself, the "*shef aviatsii*" of the Russian army. The Grand Duke and his like-minded colleagues provided both impetus and resources to further the cause of military aviation, while he possessed channels of influence and authority that most line officers could not match. As noted, the way the Grand Duke utilized this power, and his aviation record as a whole, has certainly received a mixed appraisal in the historical record. It is undeniable, however, that the Department of the Air Force and its Sevastopol Aviation School, even if measured only in terms of publicity and numbers of pilots trained, were invaluable contributors to the building of the Russian air forces. Yet, particularly as aviation expanded and became increasingly established as a service, one could argue that even if the efforts of various well-placed officers and patrons were appreciated, to many an officer in the field they served to reinforce the haphazard nature of aviation organization as well as the deficiencies of War Ministry efforts. Hence the strident calls within the press, and the plans of the War Ministry itself, to officially absorb the Sevastopol School and the Department of the Air Force itself into the army as soon as feasible.[43]

Equally or perhaps more problematic were the views of those in military circles who recognized and welcomed the promise of the airplane, yet who variously questioned the right and pretension of either the aviation institutions or the officer pilots themselves to retain exclusive say over the course of military aviation. The complaints which centered upon themes of competence, authority, and the right to determine a future path could be inverted against the aviators themselves. The assumption by the General Staff of a leading role in the direction of aviation, and the opinions of its officers who supported this policy, were the clearest expressions of such views. Yet the General Staff was not the only competing constituency with which the aviation services had to contend. In yet another article in *Razvedchik* entitled "Day of the Air Force," the author, N. Grulev, noted that "all of Petersburg" had responded to "the patriotic appeal for the strengthening of our air force."[44] Raising what he regarded as the fundamental question of the day—whether the new air force was to be composed of airplanes or dirigibles—he noted that

> by posing this question in the pages of *Razvedchik* I would like to attract the attention of broad circles of military readers to this newly arising branch of arms which, at least in the present period of its development, in no way should be considered the property of only specially military aviators. Why, before the very eyes of our century, before the astonished world of military and non-military people, a completely new branch of arms is arising, destined, perhaps, to completely alter the appearance of present war. Who but *all* other branches of arms should be receptive to this new colleague! It would be frivolous to leave it, at least in the first time of its existence, in the charge of narrow specialists, who themselves came into God's world together with newborn infants.

The article continued that this was why, in France and Germany, "all questions connected with military aviation are discussed in the general press by masses of completely ordinary readers." Regarding the relevant military authorities themselves, who were ostensibly in charge of such matters, Grulev again cited the culture of inertia and myopia that was perceived to prevail in the tsarist army: "in the monastic quiet of our military departments, the question broached by me has probably been discussed more than once; but it is absolutely impossible to wage the struggle with the aerial elements only in the chancellery cells, where there is always little air; in the meantime we must conquer that element which covers our whole planet."[45] Among both skeptics and supporters, there were those who had no wish to grant the aviators and their commanders the right to determine alone the future path of Russian military aviation.

Thus, the army as a whole functioned as audience, jury, and judge for the officers of the Russian air forces as they developed their new branch of arms and a requisite ethos of service. Within this context the growing numbers of *voennye letchiki* continued to press their case through their own initiative and example, perfecting techniques, testing equipment, experimenting with tactics, and emphasizing their own notions of duty, mission, and identity.

Ideals and Realities

In light of all these considerations which shaped recruitment, training, and service for the new cadre of military aviators, some conscious and policy driven, others more amorphous and indigenous to the *byt'* of the tsarist officer community, it is thus suggestive to ask exactly which qualities were ideally sought in a *voennyi letchik*. The lists of those recommended for the post of squadron commander from both the Aviation Section of the Officers' Aeronautics School and the Sevastopol Aviation School of the Department of the Air Force provide some clues. For example, from the list of officers finishing the Sevastopol Aviation School in Oct. 1913, four were noted as being worthy of promotion to the post of commander of aviation squadron.[46] Here are their attestations:

> -93rd Irkutsk infantry regiment Staff-Capt. I. I. Saatchi-Samadchi
> Very good military pilot. Even-tempered character. Disciplined. Good moral qualities. Well developed mentally. Treats his service duties conscientiously. Industrious. Excellent.
> -33rd artillery brigade Staff-Capt. A. I. Gornovskii
> Very good military pilot. Likes to fly. Even-tempered character. Well developed mentally. Treats his service duties conscientiously. Disciplined, Efficient.
> -26th artillery brigade Staff-Capt. A. V. Zaleskii
> Very good and talented military pilot. Flies boldly. Well-developed mentally. Good moral qualities. Disciplined. Efficient. Industrious.
> -1st Life-Guards artillery brigade Lieut. E. A. Nesterov
> Very good military pilot. Likes to fly. Even-tempered character. Mentally developed. Disciplined. Efficient. Industrious. Very good.

We have already come across several of these officers before. Despite their qualifications, three ran into difficulties for their non-Orthodox heritage, while one expressed the desire to continue his service within the Guards Corps.

In contrast, in a secret report of January 1913 report to the Commander of the Aeronautics Unit, the Commander of the Officers' Aeronautics School noted that he was forwarding the list of officers, in order of their finish, completing the Aviation Section in 1912 and designated to occupy the post of squadron commander.[47] Attached was the list of 12 officers, with short attestations, signed by Col. Ulianin. Of the twelve, six officers were judged "fully prepared" to occupy the post of squadron commander, two as "sufficiently prepared," and four as "prepared." Here are the first five:

> -Staff-Capt. Prussis, 6th Siberian sapper rifle battalion
> Outstanding pilot. Excellent technical training. Very disciplined and positive officer. Very good moral qualities. Fully prepared for the post of squadron commander.
> -Staff-Capt. Karachaev, 8th aeronautical company
> Outstanding pilot. Good technical training. Positive officer. Good moral qualities. Finished OVS in the second rank. Fully prepared for the post of squadron commander.
> -Staff-Capt. Bogdanovskii [unit not indicated]
> Good pilot. Excellent technical training. Diligent, disciplined, positive officer. Very good moral qualities. Fully prepared for the post of squadron commander.
> -Lieut. Ushakov, Vladivostok mining battalion

> Outstanding pilot. Excellent technical training. Disciplined, positive officer. Very good moral qualities. Fully prepared for the post of squadron commander.
> -Lieut. Firsov, 9th aeronautical company
> Outstanding pilot. Excellent technical training. Disciplined, positive officer. Very good moral qualities. Finished OVS in the first rank. Fully prepared for the post of squadron commander.

The two lists are suggestive, as the nuances in phrasing and emphasis attest again to the somewhat different priorities of the Aviation Section of the OVS and the Sevastopol School in training and producing pilots. Yet in many ways the most significant manifestations of divisions and tensions regarding who and who was not a *voennyi letchik* swirled around our protagonists, Mikhail Efimov and Staff-Capt. Petr Nesterov. Efimov was a pilot and flight instructor of unquestioned ability, one of the very first Russians who had learned to fly. It was due to his stature and reputation that he had been personally invited by the Grand Duke himself to assist in the noble project of building Russian military aviation. Efimov was also the grandson of a former serf and largely a self-made man. It would have been difficult to find a greater contrast between professional competence and skill and poor social pedigree. Apparently some Sevastopol officers, particularly those of an aristocratic bent, could never reconcile the former with the latter; he was resented both on the grounds of being a civilian and being a plebeian, and because he was in a position of superiority over them. While they had no choice but to defer to Efimov's knowledge and instruction in the air, once on the ground snide remarks, mocking laughter, and thinly disguised contempt were their attitude. The trainee Baron Buksgevden wrote home to Petersburg "just imagine, some peasant (*muzhik*) is lecturing us!..."[48] From the accounts, Efimov, informed of such attitudes by his lower class comrades amongst the soldier mechanics, but no doubt well aware regardless, bore these indignities stoically, convinced that the importance of his task required him to stay and do his job.[49] The numerous excellent aviators he trained and who emulated and furthered his professional ideals, such as Staff-Capt. Zemitan, were doubtless a source of necessary inspiration and satisfaction.

Efimov, of course, was not helped by his somewhat anomalous position as a civilian. Upon taking up his command Colonel Odintsov had foreseen some of these potential difficulties, worrying that some officer might insult Efimov's dignity and thus cause this most valuable resource to depart the school. He had thus petitioned the army authorities to grant Efimov an officer's commission. This request was refused; Efimov was instead rewarded for his labors with the title of "honored citizen" [*pochetnyi grazhdanin*] by decree of the Tsar.[50] Yet Col. Odintsov's worries were prescient. Efimov's questionable moral authority over many trainees and even fellow members of his staff might be judged from the fact that despite being one of the most forceful advocates of thorough technical and theoretical training for pilots, his school had a reputation for producing dilettantes, playboys, and prematurely dead aviators.

In contrast, Staff-Capt. Petr Nesterov provided perhaps the quintessential example of a *voennyi letchik* in all eyes and in all respects. Nesterov was the most outstanding figure produced by the Russian Imperial air services in the pre-war period. As an active, energetic officer and squadron commander, Nesterov continually pushed the envelope of aircraft design, performance, and tactics. A firm proponent of thorough theoretical and technical training in aviation, he again had little patience for those who regarded flying as instinct, spirit, or adventure. During his visit to the Sevastopol Aviation School in 1913, his joy at meeting Efimov was apparently tempered by the example of the "arrogant and spoiled *dvoriane-beloruchki*" who shamed Russian aviation.[51] He worked continually on aerodynamic theory and aircraft design; some of his designs eventually did receive War Ministry subsidies for testing and development.[52] Moreover, while thoroughly acquainted with technical and theoretical matters, Nesterov made a point of treating his mechanics with the dignity, respect, and consideration deserved by valuable partners. He was always a good example to his subordinates.[53] In the somewhat iconic photograph taken following his celebrated flight from Kiev to St. Petersburg, Nesterov stands before his airplane on one side of the propeller, right hand placed across his body and under his tunic, a young Napoleon of the skies; yet only two paces back, on the other side of the propeller, and accorded almost half the photograph, is his faithful mechanic Nelidov.[54] Lastly, as an ethnic Russian of the provincial nobility, historically the backbone of the tsarist regime, Nesterov seemed to represent all the best traditions of the Imperial officer community.

As they defined and asserted their place within the army, the aviation officers drew both upon well-established sources and modes of conceptualizing their status and place, while at times transforming these conceptions and infusing them with new meanings in the unique forum presented by the airplane. It is clear that in the equation of officer-pilot, the former identification, and the dictates of duty, honor, and service that accompanied it, retained powerful salience in the young but growing corps of military aviators. Again, however, this salience was manifested in disparate ways, at times exacerbating divisions within this community, at times reasserting more enduring sources of unity. Issues such as the types of tasks befitting an officer, the proper place for soldiers in the aviation services, the implications of nationality for military service, and the place of honor in relations with civilians were sources of both discord and solidarity amongst the officers of the aviation services.

In terms of social considerations, it was clear enough that a good many officers had no wish to share the skies with soldiers from the ranks, regardless of their willingness, enthusiasm, or skill. Even in the face of official policy, such officers were largely able to deploy and protect their own idea of who was fit to serve among them. Yet for those officers who did choose to make common cause with their valuable mechanics, a bond could form, even to the extent of teaching them to fly. The mutual issue—the airplane—which occupied pilot and mechanic on a daily basis, the close working confines of the schools, and the technical demands of aviation could cement a close professional relationship between officer and soldier. This is in stark contrast to the pathetic cliché in memoirs of the late Imperial army of the "reforming officer" who would try to befriend his soldiers, only to be frustrated by the seemingly unbridgeable social gulf between them, and not least by the soldiers' own efforts to preserve this gulf.

In the case of restrictions upon religion and nationality, those unfortunate officer-aviators affected were but victims of higher policies, themselves borne of broader concerns within the Russian Imperial army. Military service had traditionally been a primary mechanism in the assimilation of various nationalities; it was one of the few areas in which true equality apparently reigned amongst the Tsar's diverse subjects. Yet the army leadership's concerns regarding the loyalty and patriotism of its servitors were only increasing after 1905. Hence the elaboration of a specific nationalities policy. Yet as a result of such restrictions, officers who had freely volunteered and trained for a dangerous and uncertain career, who had been deemed competent pilots, and in some cases who had been cited as deserving of positions of command, were informed that solely due to their nationality or religion they could not be trusted to serve as they wished. Outwardly, some of those officers affected responded with expressions of deference and patriotism; others appealed to matters of fairness and due process in the application of the regulations. Yet one can only imagine the personal reactions among them. As officers they had sworn an oath to serve an Emperor who claimed to stand equally above all, yet such claims were shown to be hollow, all the more so by rules that seemed galling not just in their manifest unfairness, but in their arbitrary and haphazard application. Perhaps one can imagine an equally powerful impression made upon their colleagues who witnessed such exclusionary practices. For all involved the clash between professional, national, and imperial identities was doubtless palpable.

Despite these divisions, honor was a unifying thread which also helped provide meaning and content to service in the new community of military aviators. As was the case with their brethren throughout the army, these officers displayed and defended their honor most vigorously against real and perceived attacks on the part of civilians. Yet their sense of honor did not bear a purely negative content. Latent hostility toward the values of capitalism and the West, coupled with suspicion of the motives of civilian pilots, inventors, and enthusiasts even within the Empire, all demonstrated that these officers adhered to an ethic of duty and selflessness that they felt defined them as true, patriotic Russian servitors.

The same ethos was apparent in their relations with the broader military community. Although these relations were often a source of contention and exasperation, the enterprising efforts and progressive example of the *voennye letchiki* attracted the attention and praise of other like-minded constituencies within the army. Both *Razvedchik* and the Society of the Zealots of Military Knowledge, for example, were magnets for military professionals in the Russian army. Moreover, unlike groups such as the military engineers, lawyers, or General Staff, the military aviators seemed to offer an inclusive vision and ideal of service that would inspire and improve the army as a whole. As they demonstrated through their group flights or their work at maneuvers, it was manifestly clear that the power of the airplane would ultimately affect all within the army. Such was why the efforts of the *voennye letchiki* had received "full recognition" in the line; "daring and courage always find fitting recognition." Hence the voices of those who came forward to support the pilots and argued for "a wider road to the bold!" Moreover, through both their words and deeds these daring pilots appealed not just to their military comrades, but to a global audience, as proud Russian innovators, professionals, and officers. Demonstrating the continuing power and vitality of the Russian army, these officers found common cause and bond with many of their earthbound comrades.

There were thus many possible definitions of the proper role, duty, and identity of a *voennyi letchik*. The results resist easy categorization. For example, those officers berated in the memoirs for their aristocratic pretensions and *beloruchki* attitudes, such as a Lieut. Tunoshenskii, nevertheless appear throughout the documents leading group training flights or experimenting with new equipment and methods.[55] A Sub-Esaul Tkachev could be thoroughly professional and technically proficient while also arrogant to his comrades and remote with his servants from the ranks. Others likewise appeared to seamlessly reconcile the demands of modernization with the dictates of tradition; so it was with the Catholic officer Lieut. Shomanskii in his paean to the powerful socializing influence of the "friendly, comradely environment" of the Russian officer community. Notable too was Lieut. Nesterov, who in his service sought to combine the most novel and modern in the army—aviation service—with the most traditional and rarefied—the Guards Corps.

Nevertheless, the tensions surrounding issues of social background, nationality, honor, and military duty again illustrate that a distinct consciousness of professionalism was the ascendant ethos within the tsarist military aviation community. Such a professional consciousness did not necessarily entail disregarding or denigrating traditional norms of honor, integrity, and status. Again, the best military aviators—officers like Nesterov, Lieut. Kovan'ko, Staff-Capt. Zemitan, and Col. Ulianin—seamlessly reconciled time-honored values with new paradigms of service. However, such a consciousness did stridently demand an end to antiquated practices, obsolete values, and archaic traditions that were at once extraneous and harmful to military aviation and its imperatives of progress.

The alleged partisans of such retrograde values and outmoded traditions were again unreconstructed aviators, obsolete aeronauts, and senile higher commanders. However, memoirs and popular histories continually assert that the ire of professional aviators was ultimately directed beyond the sphere of the army itself to the very source of obsolete practices, archaic values, and Russian backwardness writ large: the regime. These works contend that it was the regime itself and its ruling classes who would not or could not find the desire, will, or resources to pursue the great national project of aviation, whether through inertia, apathy, short-sightedness, or sheer stupidity. Such dereliction inevitably led to conflict with the growing cadre of professional military aviators committed to their vision of progress, modernity, and future might.

These works frequently cite the quintessential example of this clash of cultures: the numerous crashes and subsequent deaths of Aviation Section officers at Gatchina that were caused by the pine trees which closely ringed the aerodrome. Despite numerous requests from the command of the Officers' Aeronautics School, the manager of the Court of Gatchina would not allow the trees to be cut down because the widowed Empress was accustomed to gazing upon them from her window while she reminisced. Thus, left with no choice, a group of officers on their own initiative secretly cut down the trees, while the command of the school subsequently shielded them from any retribution.[56] Whether the ancient pomp and circumstance of Imperial protocol was truly irreconcilable with the novel and urgent demands of military aviation is a subject for the next chapter.

8| Ceremony and Spectacle

Tsarist military aviation from its very origins partook of humanity's broader fascination with the culmination of its age old dreams of flight. The airplane inherently excited what J. Glenn Gray cites as the lust of the eye,[1] a lust often manifested in the mass public aviation festivals, exhibitions, and contests that were so much a part of the aviation experience in Europe and Russia in the years before the First World War.[2] This convergence between ceremony, spectacle, and military became even more pronounced within the context of the tsarist aviation community, as most of the early Russian pilots and airplanes belonged to the military, while the spirit of a noble quest was attached to the efforts of the officer aviators.

In a larger sense, parades, memorials, and holidays were a regular and integral part of army life as a whole, giving the military community the opportunity to express cherished values, celebrate accomplishments, and affirm ideals of service and conduct. So it was in the tsarist aviation service, despite the numerous cleavages examined previously. Within this context ritual and ceremony in particular functioned to emphasize and reinforce themes of community, continuity, and tradition in linking aviator, army, state, and nation. Yet, in so new an endeavor and so young a branch of arms, rituals served to create as much as maintain a shared identity. Thus, while largely drawing upon traditional trappings of ceremony, the rituals and spectacles surrounding military aviation also served to underscore the novel and unique: the concepts of duty and service which distinguished the work of military aviators, the necessary and ideal qualities of *voennye letchiki*, and their responsibility before army, state, and empire. This was especially true of the archetypal rituals involving the airplane: the spectacles celebrating the achievements of military aviation, and the rituals mourning the human price exacted by such progress.

Spectacles

In its inaugural issue, *Sevastopol Illustrated Aviation Journal* provided extensive and detailed coverage of the events marking the opening of the Sevastopol Aviation School of the Department of the Air Force.[3] On the morning of Nov. 11th the Grand Duke Alexander Mikhailovich arrived in Sevastopol for the official opening of the school. Accompanied by his personal adjutant, he was met by the Governor-General, Vice-Admiral Bostrem, and his adjutant, as well as the chairman of the council of the school, naval Captain Kedrin, and the manager of the aerodrome, Lieut. Col. Makutin. Waiting for him in one of the aerodrome's hangars, which was adorned with flags and carpets, was an assemblage of notables: Bishop Aleksei of the Tauride and Simferopol, accompanied by military clergy; the commander of Black Sea naval forces Vice-Admiral Sarnavskii; the commanders of various naval and army units; Chamberlain of his Majesty's Court Gordinskii; city head N.F. Ergopulo; and the entire staff of the aviation school, along with a detachment from the ranks. With the arrival of the Grand Duke there began a short service and consecration of the aircraft and hangars, for which Bishop Aleksei, with cross and holy water and accompanied by the Grand Duke, walked among all of the hangars.

The Grand Duke then toured the aviation field by automobile with head instructor Mikhail Efimov. By wish of the Grand Duke, the ceremonial opening of the school was completed with a photograph of the school's staff, the Grand Duke, and all those invited for the occasion. The Grand Duke then watched the flights of two of the school's instructors, Staff-Capt. Matievich-Matseevich and Lieut. Rudnev. Perhaps prophetic of the many trials and tribulations to come over the next several years, after only two minutes of flight the engine of Lieut. Rudnev's airplane stopped and he was forced to land, which he did successfully. Perhaps equally prophetic of the vistas of opportunity which lie ahead, Staff-Capt. Matievich quickly circled up to a height of 250 meters. Onlookers then suddenly noticed that the propeller had completely stopped and the noise of the engine could no longer be heard; initially seized by alarm, they then watched as he gracefully glided down to a perfect landing. The magazine account observed that

> then from all of the spectators unanimously escaped a sigh of relief and all became agitated, filled with a feeling of ecstasy before this modest hero, who in these few minutes of flight displayed such wonderful and powerful aspects of his nature and such intelligence in the piloting of the aircraft that the best aviators in the world could envy him; we should remember with pride that he is ours.

Following the ceremony a breakfast was held at noon, to which the Grand Duke invited all those present at the ceremony, the instructors, and the trainees of the school; he toasted the health of those teaching and training at the school, wishing them success and well-being in their work. Then, by request of the instructors and trainees of the school, a telegram was sent to His Imperial Highness in Tsarskoe Selo. Reproduced in the magazine, the telegram read:

> 'on the day of the opening of the aviation school of the Department of the Air Force,

> the supervisors and those who are training, moved by the feeling of boundless love And devotion to their adored Tsar, ask me to place at the feet of Your Imperial Highness their selfless readiness to work, not sparing their life, for the glory of the Great Sovereign and the fatherland. Alexander'[4]

The Tsar in his turn replied with a telegram of his own to the Grand Duke: "I thank the supervisors and those who are training at the newly opened school of aviation for their feeling of devotion; I do not doubt their readiness to work with all of their might for the good of our national [*otechestvennago*] aeronautics. Nicholas." With this auspicious display of pomp, the process of building a military aviation arm for the Russian Empire officially began.

The progress of these efforts was marked by ritual as well. *Sevastopol Illustrated Aviation Journal* again provided coverage of the first military parade in which airplanes and the staff of the school participated, on July 30, 1911.[5] The parade, at the aerodrome of the Sevastopol school, was held in honor of the birthday of the Tsarevich and Grand Duke Aleksei Nikolaevich. The hangars and buildings of the school were decked out in flags, while various line army regiments assembled in parade formation. Along the hangars were deployed fifteen airplanes, surrounded by a detachment of soldiers from the school, along with the instructors, officer trainees, and Col. Odintsov. Then Lieut. General Bebel', commander of the 13th infantry division, arrived. Met by Col. Odintsov, he greeted the soldiers and addressed the school's officers with a speech of "approximately the following contents," according to the account in the *Journal*. Like the entire parade itself, Bebel's words served to officially recognize and welcome a new yet truly blood member of the Imperial military family. Noting that this was the first time military aviators had participated in a parade, Bebel' proclaimed:

> 'warmly welcoming you, gentlemen officers of the school, headed by your commander, Col. Odintsov, with especially heightened feeling I congratulate you on your first aerial parade. The Sevastopol school is always in the eyes of the 13th division, which always gazes rapturously upon your courageous flights. On behalf of the division I attest that, selflessly conquering the air, you long ago won our deep respect and won the highly honored place befitting you, both in our ranks and in the ranks in general of our native [*rodnoi*] army. From the bottom of my heart I wish you, gentleman, further success in your heroic affair; from the bottom of my heart I wish that the enlightened and energetic leadership of the commander of the school, Col. Odintsov, would be crowned with the same further success, and I ask you to believe, that along your aerial trail there will always be raised up from up from the earth the ecstatic trembling of tens of thousands of soldiers' hearts, united in a general prayerful hope: 'May the Lord keep you and may he crown your exploits!'[6]

Gen. Bebel' concluded by noting that he would have the honor to report by telegraph to the Grand Duke Alexander Mikhailovich about the school's participation in the parade. A religious service was subsequently held, during which, despite worsening weather, the airplanes reformed; Lieut. Dybovskii, in his favorite aircraft, then took off. The service ended, more aircraft took off, doing circles in the sky above the assembled troops: "the resulting picture was extraordinary, of indescribable beauty." The regiments began to move toward the general, the soldiers in a ceremonial march; ahead of them appeared Dybovskii, "as if beginning the parade at a height of thirty meters; behind him the Farman with Lieut. Makeev and his companion, who saluted the banners and the generals; the remaining airplanes continued to do circles at various altitudes." The magazine account noted that "so ended strikingly the first parade of the Russian army with the participation of airplanes." In response to a subsequent telegram from Gen. Bebel', Gen. Baron Kaul'bars of the OVF noted that "I consider the parade a happy omen of the harmonious work of our valiant troops and daring fliers for the glory of the Tsar and Motherland."

Indeed, the work of the officer aviators was recognized and rewarded at the very highest places. On Oct. 26, 1911, twenty three officers and Mikhail Efimov of the Sevastopol Aviation School were formally presented to the Tsar in a ceremony at Livadiia. Present as well were the Grand Duke and the manager of affairs for the OVF, Capt. Fogel'. His Highness the Emperor walked among the officers, "favoring them with kind questions, thanked them for their ardent attitude, zeal, and the efforts they had exerted on behalf of Russian aviation, and wished them complete success in this new affair."[7] At the end of the presentation the Tsar took a group picture with the officers; afterwards the Grand Duke invited all those present to breakfast at his estate.

The culmination of these many months of effort came with the celebration of the one year anniversary of the Sevastopol aviation school. *Sevastopol Illustrated Aviation Journal* once again provided extensive coverage.[8] Taking some time then to reflect upon their own efforts—not coincidentally, it was the one year anniversary of the *Journal* itself—the editors used the holiday to accentuate the unity and shared commitment of the Russian aeronautical community, vowing that they would continue to endeavor to provide their readers with full coverage of the "center of Russian aviation activity," Sevastopol. They noted that upon this occasion the Emperor had deigned to bestow decorations upon the school's officers for their labors, while observing that "we can with pride announce, that at the present time we have military aviators who are not in the least inferior, in all respects, to the aviators of Western Europe."

Regarding the festivities themselves, once again the Grand Duke, his adjutant Capt. Fogel', and various invited governmental and military notables were in attendance. The Grand Duke received a report from Col. Odintsov, walked among and greeted the assembled officers of the school and the party of soldiers, and then congratulated them all on the holiday. A religious service was held, following which various toasts to the health of the Emperor, the Grand Duke, and Col. Odintsdov were exchanged; "upon the sounds of the hymn 'God save the Tsar,' a long, rapturous, incessant 'hoorah' of those present covered the toast." By request of the party of soldiers, the Grand Duke then took a group photograph with them. Following the ceremony invited guests and the officer personnel of the school went to the premises of the Naval Assembly, where the school's annual commencement exercises were conducted under the chairmanship of the Grand Duke.

Following a reading of a detailed report of the school's activity in the first year of its existence, the Grand Duke congratulated those officers who had finished "and wished them further success in the cause of aeronautics for the good of the Sovereign and Motherland." Then Col. Odintsov delivered a congratulatory speech to the graduates, which the *Journal* printed in full.[9] A proud account of the successes of the Sevastopol school, the speech was also an unabashed paean to the Grand Duke. Col. Odintsov first praised the Grand Duke's decision to transfer the remainder of the sums originally collected for reconstructing the navy to the building of an air force; reiterating a familiar argument of the OVF, Odintsov noted that this amount of money was insufficient for the construction of even one naval vessel. He added that many had protested diverting these funds to a new and unproven technology, but "the correct military opinion and the power of foresight of Our August Chairman forced the majority to agree and choose the correct path." Detailing the various results achieved by the school, Col Odintsov emphasized that

> I cannot, as a supervisor, evaluate the work of the school, but I will repeat the words of the War Minister, who has visited the school twice and observed the flights of officer-trainees: 'I heard and read much about the school, but what I saw exceeded all of my expectations; in fairness, the Sevastopol aviation school should be considered the founder of the Russian air force.'... Among the primary virtues which any military pilot should cultivate in himself should be honesty and straightforwardness, and we, officers of the permanent and temporary staff, should openly say, that only thanks to the able leadership of Our August Chairman did we achieve positive results in the performance of this new and difficult work, we were given the broadest initiative, and the military spirit—the engine of the affair—was solicitously supported by an able hand.[10]

Col. Odintsov then addressed his remarks directly to those who had benefitted most from the Grand Duke's leadership, the new officer pilots themselves. While warning that their unique path in the cause of military aviation would not be an easy one, he had nothing but complete faith in their ultimate success:

> gentlemen officers who have finished! You have put forth much effort in order to become leaders of this arm of service which has newly come into being, [and] your efforts were appreciated: on October 26 at the presentation to the Sovereign Emperor you felt that special, most gracious attention which the Supreme Leader of the Russian army wished to give you....You are now setting off to work among the troops. Of course, as apostles of a new doctrine, strong in spirit and the awareness of the rightness of the ideas you are bringing, you will meet in life much friction [and] resistance, and I believe that you will emerge the victors, having before yourselves, like a guiding star, the example of the unselfish and high-minded service to the Sovereign and the Fatherland of our August Chairman, by whose efforts the victorious spirit was instilled in the school founded by Him.[11]

Following the completion of commencement the Grand Duke invited all of those present and the staff of the school to breakfast, "which was accompanied by toasts warmly welcomed by all of those present." The magazine then printed all of the telegrams which were sent and received upon this special occasion, including again one of congratulations from the tsar himself.

With in fact a far longer history and tradition than the Sevastopol school, the Officers' Aeronautics School also employed rituals of celebration to adopt ceremonially the airplane and its servitors. At the yearly holiday of the school on July 20, 1913, the list of roughly one hundred invited guests included numerous prominent figures from the world of military aviation.[12] Among the notables were the Grand Dukes Peter Nikolaevich and Alexander Mikhailovich, War Minister Sukhomlinov and the assistant War Minister, General Shishkevich, General Zhilinskii, and member of the State Council General Polivanov. Representing the army and navy were the commanders and several assistants of the officers' cavalry, artillery, rifle, and electro-technical schools, the commanders of several sections of GUGSh, and various other ranking army and naval officers. Prominently representing the community of military aviators and aeronauts were Lieut. Col. German, Lieut. Col. Nemchenko, Col. Agapeev, Col. Odintsov, and General Baron Kaul'bars. Lastly, aviation enthusiasts included the president of the Imperial All-Russian Aeroclub and Chamberlain of His Majesty's Court Count Stenbok-Fermor, assistant chairman of the Aeroclub Count Ia. N. Rostovtsev, Prof. Zhukovskii, the civilian aviator V. A. Lebedeb, D. P. Riabushinskii, General O. E. Shtubendorf of the Imperial Academy of Sciences, Rector of Iurevskii University B. M. Sreznevskii, state councillor I. P. Shipov, the secretary of the Society of Zealots of Military Knowledge, and various other officials, professors, and esteemed citizens. In his order of the day of July 19th, the commander of the school's battalion, Col. Novitskii, noted the holiday would be celebrated by a religious service at the Il'inskii church the next day, followed by a parade conducted by the school's Commander.

The ceremonial letters sent out upon the occasion emphasized the same themes of Imperial devotion, service, and gratitude expressed in the holiday telegrams of the Sevastopol school. To the *shef aviatsii* himself, the Grand Duke Alexander Mikhailovich, Gen. Kovan'ko wrote:

> together with me the entire staff of the OVS asks YOUR IMPERIAL HIGHNESS to graciously accept, on the day of the yearly holiday of the school, sincere wishes of happiness, health, and the strength for prosperity of the national [*rodnogo*] cause for the good and glory of the dear Motherland. We lift high the goblet in toast to YOUR IMPERIAL HIGHNESS. URA![13]

Likewise, although the Officers' Aeronautics School did not possess the celebrity of a Grand Duke, its officers also received imperial recognition on various occasions. Thus, the graduating officers of the OVS and its Aviation Section were to be presented to the Tsar on July 25, 1913, at Krasnoe Selo. Also in attendance were to be Chief of the General Staff Zhilinskii, Commander of the Aeronautics Unit General Shishkevich, the commander and assistant commander of the OVS, Lieut. General Kovan'ko and Col. Uteshev, the commander and assistant commander of the Aviation Section, Lieut. Col Ulianin and Staff-Capt. Gorshkov, and the various instructors of the Aviation Section.[14]

As with their brethren in the south, the aviators of the capital also marked the concrete progress of aviation with ceremony. For example, by 1913, the rapidly expanding aviation units had only recently been organized into distinct companies. Thus, in a letter of May 1913 to the commander of the Aeronautics Unit, Lieut. Colonel German, commander of the 1st aviation company, reported the official establishment of this unit, but petitioned to have April 27th considered the date of the company's founding, as it had been functioning unofficially by then.[15] Besides this, he noted that recognizing the 27th would be "extremely desirable" considering that "this day is especially momentous for the company because on this day HIS IMPERIAL HIGHNESS SOVEREIGN EMPEROR desired to confirm the staff of the company, which in the future will provide the opportunity to specially commemorate this day in the history of the company."[16] The reply of the Aeronautics Unit informed German that his request was granted.[17]

Just a few months later, an article in the journal *Aero & Auto Life* entitled "Aerial Parade" noted that on July 18th, 1913, the first aerial parade in Russia took place at Krasnoe Selo.[18] Taking part were pilots of the 1st aviation company, with thirteen airplanes in all; they flew in three groups from Petersburg to Krasnoe Selo, the squadrons under the command of Col. German. After a tour around the field by commanding personnel, the airplanes lifted off one by one and performed a joint flight over the field, "having executed a series of complex evolutions." The pilots then set off again back to St. Petersburg, no doubt having appropriately impressed the notables in attendance.

Such formal holidays and elaborate parades were only the most prominent of similar if smaller such occasions mentioned in previous chapters. For example, in both the excursions of various chapters of the Society of Zealots of Military Knowledge to the military aerodromes and the visits of the War Minister himself to Sevastopol, spectacle, in the form of demonstration flights and performances, always played a prominent role. In other cases, such as that of Sotnik Tkachev, demonstration flights for military and civilian audiences were often a regular feature of long-distance training flights. In a similar vein, while not strictly military rituals, the public aviation exhibitions and festivals were an integral part of the development of aviation on a mass, national scale. While military participation in such events varied, they too could be used to propagandize and enlighten. Thus, July 22, 1913 was designated the "Day of the Air Force"; it was an event organized by the Department of the Air Force in several localities in St. Petersburg province in order to collect public donations. Press coverage noted that the effort was especially successful in Gatchina. About 20,000 people gathered at the military aerodrome, while the *voennye letchiki* gave the public explanations regarding the construction and piloting of aircraft. At seven in the evening flights over Gatchina commenced, in which the pilots of the Aviation Section of the OVS took part. In all sixty prolonged flights were done, with gliding landings from high altitudes. At nine the largest dirigible "Albatross" flew over from Saliuzi, and the military aeronauts also gave the public explanations of their work. The airplane flights ended only with the coming of dusk.[19]

Death

The aerial parades and holidays were certainly impressive spectacles, designed to awe and inspire onlookers and invited guests. Yet in part as a testament to the success of the airplane, by the end of 1913 military aviation had largely moved beyond the realm of novelty and spectacle into the sphere of practical work. In contrast, the deaths of aviators and the attendant funerary ceremonies were a constant and in a sense more enduring feature of the culture of Imperial military aviation. Like the holidays and spectacles, but in far more somber trappings, these rituals served to assert forcefully the unity and collective sense of devotion within the community of Russian military aviators. At the same time, these moments of catharsis acknowledged the intimate bond of this community both to the army as a whole and to the public at large, emphasizing a shared understanding of the importance and greatness of the cause of aviation.[20]

Underscoring the fine line between glory and death, spectacle and horror, that marked the progress of aviation, one of the earliest and perhaps most traumatic fatalities involved that of Captain Matsievich at the First All-Russian Festival of Aeronautics. As noted in chapter two, this incident provoked a mass outpouring of public sympathy and mourning, replete with a grand state funeral and a memorial to the deceased.[21] Yet the fledgling Russian military aviation community received another shock just several months later with the loss of Staff-Captain B. V. Matievich-Matseevich and his brother, Warrant Officer S. V. Matievich-Matseevich, killed in a crash at the Sevastopol Aviation School on April 18, 1911. This event, especially disturbing as it was, also received extensive coverage in *Sevastopol Illustrated Aviation Journal*.[22] Described as "the best pilot of our aviation school," the editors lamented that "the pride of our air force," a pilot who "aroused a storm of ecstasy by his flights during the aviation week in St. Petersburg," had been lost forever. They reminisced that Matyevich's very first flight on the day of the school's opening "immediately showed what great power we have in our battle with the aerial ocean." They added that he was a teacher as well, "a talented, persistent, beloved teacher," who had already trained such talented aviators as Lieut. Dybovskii and Lieut. Samoilo. The editors thus intoned that "all citizens of Sevastopol, all of Russia join the grief" of the family and friends of the Matievich brothers; they noted somberly that on the field of the aerodrome "has been shed new blood, the blood of the best sons of our motherland." Nevertheless, the editors assured their readers that "to their earthly remains all of Russia gives their regards and marks their names deep in their hearts. *May their memory live forever!*"

Regarding the details of the crash, the *Journal* reported that the aircraft had apparently been hit by a gust of wind during flight, which tilted it first to the right and downward; at this point the plane ceased responding to the wheel, continued heading down, and then smashed forcefully into a stone wall, killing both pilot and passenger. The subsequent investigation determined that the probable cause of the accident was that a metal cable which stabilized the wings had worn through, as a result of which the pilot lost control of the airplane. The fact that those present at the aerodrome did not hear the engine merely confirmed that Matievich, "as an experienced flier, who in many cases of his aviator activity did not panic, here, in the last minute of his life, did not become flustered." Instead, trying to correct the free fall of the airplane and seeing the oncoming wall, he cut the engine and tried to glide to a landing in front of it, but there was apparently not enough room.

The *Journal* included a telegram the Grand Duke Alexander Mikhailovich had dispatched in response to the accident: "deeply shocked and sincerely mourning at heart, I share with the school our great common grief; but I am convinced that this misfortune will not demean the glorious spirit of all of the personnel of the school." The account then provided extensive details of various ceremonies to mark the passing of these officers. Those in attendance at the first funeral service included the by now familiar senior naval officers of the Black Sea fleet; Lieut. General Bebel'; personnel and instructors of the Sevastopol Aviation School, headed by Colonel Odintsov; and assorted other commanding officers. Another service that evening attracted other senior commanders, the entire personnel of the Sevastopol School, and "a mass of worshipers, filling the chapel" and the surrounding area. Wreaths from the 13th artillery brigade, the 50th Belostok regiment, the Sevastopol sappers, the military engineers of the Sevastopol fortress, and the editors of the *Journal* were laid at the coffin.[23] On April 20th the body was moved from the naval infirmary to the Catholic church, "accompanied by the entire staff of the school and a huge mass of city dwellers." A mass was held the next day, while the funeral procession was to be held on April 22. That same evening the body of Matyevich was to be transported to Zhitomir, to be buried in a common crypt with his parents.

The *Journal* then printed the telegrams of condolence sent to the commander of the Sevastopol Aviation School by various members of the Russian military aviation community and concerned sympathizers.[24] To note just a few, Major-General Kovan'ko of the Officers' Aeronautics School telegrammed:

> '[we are] deeply saddened by the tragic death of Staff-Capt. Matievich-Matseevich. Today in the school church a funeral service was conducted in the presence of the commander of the Main Engineering Administration and the officers and soldiers of the school. Lieut. Rudnev was appointed as the school representative for the laying of a wreath at the coffin of the deceased.'

From St. Petersburg, the aviator Lieut. Piotrovskii telegrammed: "I am shocked by what has happened. I give my regards to the earthly remains of a glorious colleague." Also from St. Petersburg, Col. Nemchenko telegrammed: "deeply shocked by the death of the courageous officer-pilot Matievich-Matseevich, the Committee of the Imperial Aeroclub expresses feelings of the most heartfelt condolences to the school." As noted by the editors of the *Journal*, the death of Matyevich not surprisingly made a "strong impression" on the staff of the school. As they made clear, however, echoing the Grand Duke's initial telegram,

> the duty of service, a duty taken upon themselves by the instructors and officers training in the cause of the conquest of the air, prevailed. The tragic occurrence of the death of a colleague and teacher was taken by the staff of the school as a casualty of wartime. 'We are in a war, a great worldwide war, with the conquest of the air, and there cannot be a war without casualties' was heard among the ranks of the staff of the school. And this phrase of the staff of the school was fully confirmed by the daily calm continuation of training, and the investigation into the reasons for the accident is being engaged in with the same composure.

The article noted that this attitude was confirmed by the commander of the school himself, Col. Odintsov, in his order of the day:

> 'assembled by me today, the officers, having listened to the telegram from the August Chairman of the O.V.F., expressed to me their adamant intention to persistently continue their usual work, with the aim of achieving the fondest goal- victory over the air, for the strengthening of the power of the Sovereign and motherland. As commander, I am proud of that truly great spirit, which reigns in the souls of the officers who are training. A fighting man who has been put out of action will be replaced by another. Those who have entrusted this new affair to us can be comfortable, that temporary failures and misfortunes, inevitable in life, will not affect the victorious spirit of the Russian pilot.'

The *Journal* duly noted that on the 19th the entire staff of the school gathered at the usual time for training, but due to strong wind no flights were conducted.

By a decree of an assembly of the officers' society and with the agreement of the school's commander, another funeral service was held that morning at the spot of the brothers' death, and a wooden cross planted. Lastly, the editors reported that there was a proposal to request permission from the city administration to place a memorial at the spot of the crash, "if only in the form of a column." The Grand Duke Alexander Mikhailovich duly arrived in Sevastopol on April 22, where at two that afternoon he visited with the widow of Staff-Capt. Matievich, Maria Alekseevna, and "with deeply felt words" consoled her in her grief. At three he went to the school, where he addressed the entire assembled staff; he noted that he and the Emperor, whom he had several times informed of their successful work, thanked them for their efforts and progress.

Matievich's death reverberated far beyond the environs of Sevastopol. While commenting upon the International Aeronautical Exhibition being held in Moscow, Aleksei Poltaratskii, again a regular contributor to *Razvedchik* on aeronautical affairs, reminisced proudly about Matyevich.[25] In a subsequent report from the Exhibition, Poltaratskii noted that while the assembled public waited anxiously for the appearance of the Frenchman Bleriot himself to deliver a lecture on aeronautics, Col. Naidenov appeared. Informing them that the lecture would begin after the first part of a scheduled concert, Naidenov proposed to honor the memory of the dead Matievich-Matseevich by rising; a funeral march followed. Poltaratskii observed that "a cold wisp passed over the auditorium—the freezing wind from the black wings of death... Among those present were many witnesses to the death of Captain Matsievich, and what happened in Sevastopol appeared so clearly to them...." Subsequently, Poltaratskii noted, no one listened to the concert.[26]

The same grim litany of events was repeated at the Sevastopol School upon the crash and death on Jan. 26, 1912 of Staff-Captain F. F. Leon of the 17th mounted artillery brigade and Staff-Captain V. N. Ianish of the Siberian Aeronautical Battalion, approximately three minutes after takeoff. As always, *Sevastopol Illustrated Aviation Journal* provided extensive coverage of the accident itself and the subsequent ceremonies.[27] After relating the ugly details of the crash, the account reported that in what was also becoming a regular and unpleasant ritual, the council of the OVF met to ascertain the reasons for the accident. After determining that both aircraft and engine were in fine working order, while the weather was calm and clear, the council concluded that the plane had been hit by a gust of wind. Staff-Capt. Leon, attempting to correct the resulting bank, made a mistake in steering, taking too sharp an incline downward, while the low altitude did not give him time to correct this or shut off the engine, which strengthened the crash impact.

As was becoming custom as well, the commander of the school received various telegrams of condolence from concerned members of the aviation community and the military as a whole. The Grand Duke Alexander sent word that "I grieve deeply over the dead officers." Baron Korn telegrammed that "the Imperial All-Russian Aeroclub, deeply shocked by the mournful news from Sevastopol, sincerely mourns the grievous loss which has befallen the Russian aeronautical family." Lieut. Piotrovskii sent word that "with sorrow I learn of the untimely death of [my] dear comrades, I bow low to their remains." Lieut. Gel'gar, a *voennyi letchik*, telegrammed to say that "I am depressed by the untimely death of [my] dear comrades," while Lieut. Gausman, another *voennyi letchik*, sent word that "deeply stricken by the death of dear comrades, I sincerely join the general grief of [my] home school." Lastly, the commander and staff of the warship *Ioann Zlatoust*, "deeply shocked by the tragic death of the valiant aviators," also telegrammed the school to express to "their sincere sympathy for the sorrowful loss."[28]

Coverage in the *Journal* continued with the accounts of the religious services. The two services in the chapel of the 50th Belostok regiment were attended by all of the personnel of the school, officers of the Belostok regiment, senior commanders, the families of the pilots, and many of those who knew them. The bodies were moved to the naval church of Vladimir the next day. In attendance were the commander of the 1st brigade of the 13th infantry division, Maj. General Lavren'tev, and a platoon of the 8th aeronautical company under two officers. After a short litany soldiers of the school and the 8th company removed the wreaths, while the officers carried cushions with the medals of the deceased; the coffins followed, carried by Col. Odintsov, Capt. Prince Muruzi, and officers of the school and the 8th company. They were placed in chariots waiting at the barracks gates, pulled by two horses clad in white mourning vestments; there followed a "mournful procession" to the Church of Vladimir, where the bodies were to remain until burial.[29] Services there were again attended by the familiar high-ranking officers of the Sevastopol garrison and naval forces, various army and naval officers, members of the pilots' families, and numerous mourners. In all 37 wreaths were laid at the coffin, among them two from the August Chairman, others from the comrade aeronauts of the OVS, the Sevastopol school, the officers of the 8th aeronautical company, numerous army units stationed in and around Sevastopol, the city administration, various leading figures in the city, and the editors of the *Journal*.

As was the case with Matievich, the *Journal* concluded its account by reporting that following the crash, on the very same day, the officers of the Sevastopol school gathered for their usual exercises at about three in the afternoon. Despite the terrible accident, the wreckage of which could still be seen, "the duty of service, the consciousness of the responsibilities taken upon themselves and their responsibility before the state ordered them to begin their usual exercises. One after another the airplanes began to rise into the air." The instructors Zemitan and Andreadi made a total of three flights over the city.[30]

The following issue of *Sevastopol Illustrated Aviation Journal* continued with extensive coverage of the burial of the officers on Jan. 29th.[31] A deputation of four officers arrived from Staff-Capt. Leon's regular unit, the 17th mounted artillery brigade, and presented a wreath from their comrades in arms. At nine in the morning a "mass of people" gathered to pay their last respects; the huge church could not hold all of those who had come. The family arrived, as did a deputation from various military units, "the entire *generalitet* of the Sevastopol garrison," headed by Vice-Admiral Sarnavskii, and many other naval and army officers. Following the liturgy the coffin was carried out in the hands of the commanders, pilots of the school, and officers from the 8th aeronautical company. The ensuing procession, "accompanied by a huge crowd," wound through the streets and then across the anchorage. Along the entire route stood the troops of the Sevastopol garrison, with arms and choruses of music, while on the ships at anchor parties of sailors and officers stood in full parade dress as the numerous launches of the procession passed by. Once on shore the coffins were placed on a gun carriage and moved on to the military cemetery. There the final funeral service was held, and the remains of the officers were laid to rest in a common grave amid volleys from the 8th aeronautical company and the field artillery. Immediately afterwards a cross made of two propellers was raised on the grave.

The *Journal* then printed the eulogies which had been delivered during the funeral services that day. The first was from the priest R. Medved, who had also officiated at the funeral of Matievich. In more elaborate form, he reiterated to his audience the sentiments expressed by Col. Odintsov in his order of the day following the death of Matievich. He first exhorted them to ignore those who pointed to the almost daily victims of aviation and the danger involved. He noted that aviators knew well what they were seeking, but wondered aloud whether those present knew the full worth of the dead pilots' feats:

> after all, aviators traverse their dangerous path not for personal pleasure, but for the glory and honor of the fatherland. And we ourselves are this fatherland. We are living members of the fatherland, and that which is done for the motherland is done specifically for each of us. At the sight of these sacrifices of love of the fatherland we are obliged to know, that these sacrifices were made for us.[32]

The priest noted that the full church testified to the fact that those present were not indifferent to these sacrifices, and assured them that before the Divine Judge, the deceased would no doubt receive "full requital" for their sacrifice, "made by them for the glory of the fatherland." He reminded those assembled that the greatest feat was to sacrifice one's life in the name of Christ; after this one of the loftiest feats was the sacrifice of one's life for one's motherland." He thus prayed that God, taking into account their feat, would forgive them any sins, and that Vladimir and Fedor would "immediately enter the eternal kingdom of the saints."

The *Journal* then "approximately" reproduced the words of the improvised eulogy delivered by the priest of the Black Sea fleet, Fedor Milianovskii. The account noted that his fleeting acquaintance with Staff-Capt. Ianish had nevertheless made a strong impression on the priest, which was one of the compelling reasons for his speech. The priest explained that the desire to rise above the earth was characteristic of humanity's need to search, to quest, without which "mankind would spiritually die." The conquest of the air was perhaps the most difficult forms of humanity's efforts to possess nature's secrets, and "not everyone finds in themselves such courage to undertake this cause." Those serving this cause did so for the common good, and those who perished doing so were especially precious to all:

> in their person we see and revere the glorious Russian warriors, the heroes of the Russian spirit, Russian might, Russian genius, and however grievous, however bitter their loss is for us, their death lifts in us faith in the hardiness of the Russian people, which gives the motherland sons so strong in spirit and glorious in their exploits; their death shows that fearlessness has not been exhausted in the Russian warrior, that the Russian warrior is selflessly devoted to his duty, that the Russian warrior does not fear death, that he, when it is necessary, not hesitating, will place his soul in its hands.[33]

The priest added that the death of "our warrior-heroes" was mourned by "all of our compatriots" and the motherland as well. Emphasizing that "immortality belongs to the brave," he blessed the coffins and wished "immortal and glorious memory" to the dead. In conclusion, the priest addressed all present: "'Orthodox Russian people! Bow down to the earth before the best of our sons and with love bid them a last farewell.'" Following a requiem service at the Bratskii cemetery in Sevastopol, the memorial erected at the grave of Staff-Captains Leon and Ianish was consecrated on Jan. 26, 1913, the one year anniversary of the crash. Those present at the ceremony included unit commanders, the commanding and officer staff of the school, and the public. The memorial consisted of a column on a granite pedestal with a propeller on the top, under which was noted, in golden letters on a marble tile, the date of the "heroic death" of the pilots.

Like the aviation spectacles, even funerals eventually lost some of their power to shock and attract mass public attention. Yet deaths and the rituals attending them still performed much the same function: to eulogize, to enlighten, to unify. As aviation crashes increased alarmingly throughout 1912, 1913, and 1914, the journal *Vozdukhoplavatel'* provided a series of obituaries for the victims. In nearly all them the dead pilots were presented as all that was best in both aviator and Russian officer. This is not surprising, as the journal was edited by Lieut. Col. German, longtime officer of the Officers' Aeronautics School (the masthead, in fact, merely listed the OVS as editor) and had various officer pilots as regular contributors.

Thus, the journal devoted extensive coverage to the death of the *voennyi letchik* Second Lieut. A. P. Perlovskii of the Brest-Litovsk Aeronautical Battalion on March 15, 1913.[34] Flying on a Nieuport above Warsaw, he had fallen and crashed from a height of about 150 meters. The account of his death, penned by an unidentified colleague who had known him, spoke in glowing tones about this fallen comrade. Perlovskii had begun his service in the 4th Siberian sapper battalion, and "immediately called attention to himself as an outstanding officer; with youthful ardor he set about his work and performed the most diverse assignments distinctly and brilliantly." From childhood he had had an enthusiasm for all manner of sports, and "for a long time burned with the desire to familiarize himself practically with aeronautical affairs." Thus, after serving the minimum time necessary for entry into the Officers' Aeronautics School, he sought entry and was enrolled in September 1911. Once at the school Perlovskii "won himself great respect and sincere love among comrades and commanders. An excellent comrade, always cheerful, buoyant, the deceased infected all with his gaiety." Further, Perlovskii "became fully engrossed in aeronautics and was devoted to the affair in the full sense of the word, to the point of self-disregard." The account noted that during practical exercises OVS officers were required to work from 7–9 hours a day, with a two hour break, which most officers, so very tired, devoted exclusively to rest. Yet, "rest A. P. did not know; flawlessly performing all that service demanded of him, he devoted his free time to training in the aviation school of the Imperial All-Russian Aeroclub." Already on July 11, 1912, he had passed his pilot's test at the Aeroclub.

Perlovskii did not limit his enthusiasm to aviation: "as a person richly gifted and with a broad range of interests, he loved aeronautics in its full scope." The account noted that during the "endless arguments" in which his fellow officers would argue over the superiority of different aircraft systems and their desire to fly them, he usually shrugged his shoulders and said 'I don't understand what you are arguing about, gentlemen. Believe me, for us, in our role, every aircraft is precious and invaluable—they give us, students, practice being in the air, and one must prize every minute spent in the air.' The piece noted that Perlovskii confirmed his words with deeds, spending the most time in the air during exercises out of the group of 37 officers training in 1912. Perlovskii, "known for his outstanding aptitude for aviation" was then assigned to the OVS in November to take the test for the title of *voennyi letchik.* He passed and remained at the school to train on Nieuports, "of which he had dreamed for a long time." The piece sadly concluded that

> on March 15 God sent a grievous trial for Russian military aviation—Second Lieut. Perlovskii perished by glorious death in the performance of his duty, having become the first victim of the Aviation Section of the Officers' Aeronautics School. May you rest in peace, our dear comrade who untimely perished, the blessed memory of you will remain forever in the hearts of all those who knew you.[35]

It is worth noting that Perlovskii's death was noted in far less sentimental fashion in the report of the Main Administration of the General Staff which complained about the level of training provided at the aviation school of the Imperial All-Russian Aeroclub.

The following issue of *Vozdukhoplavatel'* brought more grievous news.[36] It concerned the crash and death in New Peterhof on May 18, 1913, of the *voennyi letchik* Lieut. M. G. Balabushka.[37] After serving a year in the 11th sapper battalion, he had entered the Officers' Aeronautics School and completed the course in outstanding fashion, earning a medal decoration. Posted to the 11th aeronautical company and then to the Aviation Section of the OVS in April 1912, he again served in exemplary fashion and earned another medal. Transferred to the newly formed 1st aviation company, Balabushka "with his characteristic energy" set about his new duties, "but at the very beginning of his work premature death took him to the grave." The account attempted to describe the sum total of the loss:

> in the person of M. G. Balabushka his aged mother lost an ardently loved and loving son, who, after the death of his father, helped her; the Russian army lost an outstanding officer, a brave, energetic *voennyi letchik*; officer-pilots—a glorious, kindhearted comrade, who by his zest for life and gaiety, even in difficult moments, was able to instill in their souls joy and energy.[38]

The magazine then provided an account of the funeral and burial of Balabushka at Volkovo cemetery on May 21. Present were all of the officers of the 1st aviation company, headed by the company commander, Lieut. Col. German, as well as most of the officers of the 148th Kaspisskii infantry regiment, together with their commander. The procession wound its way across all of Peterhof to the church of the 148th regiment, where a funeral service was performed. A train then took the body to the Baltic station in Petersburg; waiting to meet it was Chief of the General Staff Zhilinskii, Lieut. Gen. Kovan'ko, Lieut. Col. Ulianin, the general-quartermaster of the Petersburg military district, Maj. Gen. Erdeli, and many others. After a short liturgy, the funeral procession headed to the Volkovo cemetery. A Nieuport appeared in the air above the procession, flown by Lieut. Pavlenko of the 1st aviation company, "who from the heights of the clouds greeted his perished comrade in arms." Balabushka was put to rest in the same place as the four victims of a flight over the Gulf of Finland in 1907. On his fresh grave a cross made of airplane propellers was erected.

The by now ritual condolence telegrams came from a range of the various aviation units. Many of them were sent, as one telegram openly noted, from comrades who had trained at Gatchina:

1) "We deeply pity our perished comrade, lay a wreath with the inscription- from the pilots and aeronauts of the 8th company." Likhachev (Sevastopol)
2) "We ask that a wreath be laid for Lieut. Balabushka with the inscription 'from the comrade Gatchintsy of the Kiev squadron.'" Lieut. Ivkov (Kiev)
3) "To our dear Balabushka, our dear, glorious comrade, I send my last regards." Nikiforov (Brest-Litovsk).
4) "The entire aviation squadron with all of its heart shares your grief over the untimely death of a dear colleague, may he rest in peace.' Lieut. Col. Liakhov (Osovets)
5) "The Kovensk pilots deeply mourn the death of a dear comrade." Lakhtionov, Sheremetevskii, Popov
6) "The 3rd aviation company expresses its condolences over the grievous loss of a comrade-pilot, we ask you to lay a wreath." Lieut. Col. Solov'ev (Kiev)
7) "The IMPERIAL Aeroclub, deeply grieving over the untimely loss of Lieut. Balabushka, expresses its sincere sympathy to you; I extremely regret, that this urgent matter does not allow to attend the funeral.' Chairman Count Stenbok-Fermor (St. Petersburg)
8) "We deeply sympathize with the grief which has befallen [you]." Officers of the aviation squadron of the 5th company (Grodno)
9) "The officers and soldiers of the company sympathize with the grief of [our] comrades, lay a wreath." Capt. Fetisov (Grodno).[39]

Several months later, *Vozdukhoplavatel'* reported on the consecration of the monument to Lieut. Balabushka on Oct. 1st at New Peterhof. Located on the parade ground of the 148th Kaspiiskii regiment, the "modest" monument consisted of a large rock with a square inlaid tile, on which was carved out an inscription marking the date of the crash and Balabushka's name, rank, and unit. Those in attendance at the ceremony were officers of the 148th regiment, headed by their commander, Col. Shil'dbakh, the commander of the 1st aviation company, Lieut. Col. German, and many officers of the company.[40]

In a somewhat fitting if unfortunate coda to this procession of perished aviators, the regular "Chronicle" feature of *Aero and Auto Life* reported on the crash and death of the renowned Staff-Capt. Andreadi, instructor at the Sevastopol Aviation School, on March 7th, 1914.[41] Active in the Sevastopol School from its inception, it was Andreadi's daring and successful flight that helped convince the Grand Duke to select modified Nieuports to stock the aviation squadrons. Yet for some reason Andreadi's airplane had banked sharply in good weather; unable to level it, he and the airplane fell from a height of about 100 meters. The account noted that he was the 30th victim of Russian aviation; moreover, news of the death of one of the best Russian military aviators "produced a dispiriting impression in aeronautical circles." Great hopes had been placed on his recent work in aircraft construction; with this is mind, at the end of the previous year Andreadi had attended the course in theoretical aviation at the St. Petersburg Polytechnical Institute.

Once again, telegrams of condolence were received from various aeronautical organizations and military units, while Andreadi was to be buried in his native Simferopol. The account then recounted an interview published in *Vechernee vremia* with Lieut. Koltovskii, a comrade of Andreadi. Koltovskii observed that "from the very first day" Andreadi had distinguished himself "as a very talented and daring pilot." Trained first on Farmans, he had switched to Nieuports, which was his chief interest; Andreadi piloted this craft "marvelously;" "on a 'Nieuport' he truly felt like a bird in the air." As an instructor, Koltovskii continued, Andreadi was notable for being very conscientious: "some of his students even expressed dissatisfaction, complaining that he was holding them back and was not allowing them on independent flights." "Extremely careful in regard to others," Andreadi would reply in such cases: "'you will fly better later, but calmly and confidently.'" Koltovskii recounted one incident where Andreadi's grief over a fatality involving a student was matched only by his reproaches of the student's reckless mistake. In turn, Koltovskii noted that Andreadi had been told several times that he himself was too careless, too daring. He quoted Andreadi's alleged reply: "'you know, I am so sure that nothing bad will happen with me in the air, that I literally cannot even imagine such a situation from which I could not find an exit.'" Koltovskii marveled that Andreadi "performed such stunts, that we, pilots, unintentionally closed out eyes. For him the weather completely did not exist." Ultimately, however, Andreadi's confidence, skill, and daring could not save him; he would not live to see the fruits of his and his comrades' labors.

Ceremonies of both celebration and lamentation, through united choruses of "hurrah" or the common grief of mourning, in turn reinforced by related displays of unity and support, expressed in palpable and visceral form a common sense of purpose and shared ethos within the Russian aviation community. However, such rituals could also serve to illustrate and even accentuate the professional divisions that persisted among the ranks of aviators and officers. Despite the paeans paid to dead heroes and the trappings of funerary rituals, aviation crashes in particular often provoked commentary amongst both officers and sympathetic observers regarding the technical, professional, and personal factors at fault in such sad occurrences. Such was the case, for example, with the death of Lieut. G.I. Polikarpov of the 1st aviation company.[42] Called by *Vozdukhoplavatel'* "one of the outstanding" *voennye letchiki*, Lieut. Polikarpov was a 1912 graduate of the Sevastopol Aviation School. He was killed in a crash on July 23, 1913, along with a mechanic of the Russo-Baltic Factory, Drenov, while flying a Sikorskii aircraft from Petersburg to Krasnoe Selo. Once again unity and homage were the rule. Polikarpov was described as an "extremely modest, likeable person; all of his comrades liked him. He was very interested in aviation."[43] At the funeral of the two men on July 25th, those in attendance included Gen. Zhilinskii, Gen. Shishkevich, Chief of Staff of the Petersburg military district and member of His Highness' Suite Maj. Gen. Gulevich, and others. More than thirty wreaths were laid upon the grave, including one from the August Commander in Chief, Grand Duke Nicholas Nikolaevich. The dead were laid in a common grave at the Krasnoe Selo military cemetery.

However, in the same issue of *Vozdukhoplavatel'* that reported Polikarpov's death, the editors addressed an uncomplimentary account of the incident printed in—where else?—*Vechernee vremia*.[44] They noted that the *Vechernee vremia* piece had apparently been penned by an anonymous pilot—he signed his article "V. L.," for *voennyi letchik*—"who had taken upon himself the 'worthy' role of a person as if specially invited by the editorial staff for the discrediting of the activity of the higher administration of aeronautical affairs in our army in the eyes of the public." *Vozdukhoplavatel'* then proudly reprinted a rebuke to this piece which had recently been published in the official military newspaper *Russkii invalid*. The *Russkii invalid* piece noted that the anonymous pilot of *Vechernee vremia* first reproached the higher command for the fact that Lieut. Polikarpov had been flying an aircraft which had still not been procured by the War Department, and then raised the supposition whether this flight had in fact been done on the orders of the Aeronautics Unit of the General Staff. *Russkii invalid* responded that Lieut. Polikarpov had himself requested permission to train on this aircraft, while there was nothing at all unusual about this; if the anonymous author was really a *voennyi letchik*, it was noted, he would have known this.[45] Further, the 1st aviation company was subordinated directly to the chief of staff of the Petersburg military district, which was the only organ from which it could receive orders. *Russkii invalid* thus concluded, in words echoing Gen. Kovan'ko's previous complaints, that the item by the '*voennyi letchik,*' "is already not the first, and it is impossible not to come to the conclusion that the general tone of his items on aeronautics, striving for the discrediting of the activity of the higher administration of aeronautical affairs in the army, shows that hidden in their writing are goals which have nothing in common with the good of aeronautics."[46]

An almost treasonous piece printed by a known enemy of the military aeronautical establishment was bound to provoke a defensive and indignant response. Yet this was not the only source to raise such questions. For example, while not military aviators per se, the editors of *Sevastopol Illustrated Aviation Journal* were intimately familiar with the work, routine, and personnel of the Sevastopol Aviation School. In their commentary on the deaths of Staff-Capts. Leon and Ianish, they noted that every incident, as a first duty, provoked "the question—why?" Suggesting pilot error, the editors added that it was "interesting" in this case that the choice of who would fly was decided by lot, noting for the second time in their coverage that as a result "the more cool and perhaps even more experienced" Ianish became the passenger. They then surmised that if one presumed a mistake on the part of the pilot Leon, the presence of Ianish might have allowed him to provide useful advice to Leon at the decisive moment, provided there was enough time. However, echoing in a fashion the concerns raised by General Staff Col. Bulgarin regarding the difficulties in training observers as pilots, the editors continued that this scenario might also lead one to believe that Ianish, seeing Leon's mistake, hurried to help him not with advice, but with action. The result would have been two pilots were trying to pilot the plane simultaneously, which lead to the accident. They noted that "during the flight of an experienced pilot with a student this could never be." Nevertheless, the question why the crash occurred, "however regrettable, however painful," remained open.[47]

Such implied criticisms were relatively mild and largely deferential, generally sharing in the common grief. More damning criticism came from closer quarters. Thus, an article entitled "Glory to perished heroes!" which appeared on the very eve of the war provided extensive commentary on the recent growing numbers of casualties in the air.[48] Noting that while the instances of the aircraft or its parts breaking in the air were relatively rare, "in particular many aviators have perished" due to pilot error. The piece noted that even experienced pilots, like Staff-Capt. Andreadi and Staff-Capt. Stoiakin, had fallen victim to what it called "exactly their unfamiliarity with the correct piloting of airplanes." As evidence, the article cited a lecture delivered to the Moscow Society of Aeronautics on July 10 by the noted *voennyi letchik* Staff-Capt. Petr Nesterov, in which he addressed the issue of the recent losses of Russian pilots. The article emphasized that in his report Nesterov proved that almost all dead Russian aviators had fallen victim to the "obsolete method of training in piloting." Nesterov considered the fundamental mistake in contemporary training to be the "surprising conservatism in the techniques of piloting. What was taught in 1909 is still taught now, while many of the fundamental techniques are incorrect." For example, the common practice of descending into a turn—Nesterov noted one should in fact do the opposite—was the reason for the death of such an experienced pilot as the renowned Andreadi.

The article's author noted that he could not help but agree with Nesterov's interpretation and conclusions, arguing that it was necessary to reform the entire system of training. He punctuated his appeals with sentiments no doubt shared by the studious and erudite Nesterov: "indeed, aviation schools should turn out students fully armed with knowledge and experience in the struggle with the aerial elements, as only then can they be of use to aviation. Otherwise, all of their heroism will be only 'the folly of the brave.'" The author reiterated that it was incumbent upon the aviators to always remember the danger of their cause, the treachery of their enemy, and their as yet incomplete victory over the air: "then, being constantly prepared for attack, aviators will be better able to repulse their enemy, and by this preserve many courageous heroes for their fatherland!"[49] Likewise, Staff-Capt. Nesterov no doubt mourned his fallen comrades as deeply as any officer, yet this did not stop him from expounding upon the needlessness of their deaths. Far from a noble sacrifice, fatalities were the useless and avoidable result of obsolete training methods.

As such, the meaning and legacy of ritual was both fluid and relative. Somewhere in between the vocal unity and outright dissonance provoked by rituals was ambiguity. For example, numerous memoirs of officers, government ministers, and court officials regarding presentations to the Tsar attest to the fact that many of those being honored were invariably underwhelmed and disappointed, not awed, by the Sovereign yet socially awkward Emperor. In another sense, the presentation to the Tsar of the officers of the staff of the Sevastopol aviation school in 1911 is an incident invariably cited with disgust in Soviet memoirs and historiography.[50] While seemingly an occasion to celebrate, it placed Mikhail Efimov, instructor and civilian, in a delicate situation. He was the only non-officer participating, and unlike them possessed no formal parade uniform or such. Choosing what to wear became a matter of some difficulty, and his predicament was thus a rather awkward one. Accordingly, he was mocked both before and afterward by his aristocratic nemeses among the school's staff, while even those officers well-disposed to him apparently could not resist a chuckle over this particularly embarrassing situation. A small incident, true, but was this a positive affirmation of a distinctly military identity, or a testament to their lack of professional respect as aviators for Efimov?

A weightier issue concerned how ritual functioned in light of the division between the two centers of Russian aviation, as well as within the larger context of the officer corps as a whole. A common assertion, at the time and in retrospect, was that a multitude of minor identities and loyalties—to school, regiment, branch of service, and on and on—precluded any overarching sense of esprit de corps for the officer cadres. Thus, what should be made of Col. Odintsov's 1911 speech to the graduating officers of the Sevastopol Aviation

School in which he continually cited the unique example of the school and even quoted the War Minister that it should be considered the founder of the Russian air force? Was this merely a justified and appropriate display of pride in their accomplishments, particularly when the Officers' Aeronautics School was still largely organizing its own aviation efforts? Or could this be construed as promoting competition and even divisiveness? A clue might be provided by Gen. Kovan'ko, who in correspondence with Col. Ulianin in 1912 fretted that OVS officers being presented to the tsar should in no way be placed at a disadvantage before their Sevastopol counterparts.

Likewise, what of the predilection of *Vozdukhoplavatel'* to eulogize its "own" Officers' Aeronautic School officers? The deaths of Sevastopol aviators like Polikarpov were covered by the magazine, but generally not with the depth of feeling and detail reserved for OVS graduates. In a similar vein, what should be made of the telegram dispatched upon the death of Lieut. Balabushka from his "comrade Gatchintsy" of the Kiev aviation squadron? Regarding another such telegram, Col. Likhachev, despite being lambasted by Col. Muruzi for his "incomprehensible attitude" and prejudice against aviation, sent a comradely message of condolence upon the death of Balabushka. On an occasion of shared grief and expressions of solidarity, were these positive affirmations of a common experience or a divisive reminder of the differences between the two schools?

Despite the ambiguities and dissonance manifested in such responses, it is easier to elucidate the values that those presiding over these aviation rituals tried to inculcate, the lessons they sought to teach, and the deeper truths they wished to impart. Most prominent in this regard were the themes of excellence, tradition, continuity, and community. What strikes one first is the consistent message of the funeral rituals in particular. The dead aviators who were lamented in speeches, telegrams, and obituaries were presented as emblematic of all that was ideal in a *voennyi letchik*—personally, professionally, patriotically—and thus by their example transcended real life cleavages within the cadre of military aviators. Thus, such officers were notable for their initiative and enthusiasm, both in their pre-aviation careers, which frequently did not provide sufficient outlet for their talents, and then as trainees and aviators. They were rugged sportsmen, yet they were also serious about and committed to their interest in aviation. They were skilled and conscientious pilots, yet also brave and daring, sometimes to a fault. Lastly, they were of course fine and trusted colleagues, who enlivened their comrades by their gaiety and motivated them by their example. The ideal presented was powerful and persuasive; against it dissent was difficult.

Equally striking is how this ideal was firmly situated within the community and traditions of the officer corps as a whole. Specifically professional considerations were increasingly distinguishing the *voennye letchiki* from the engineering corps or even their fellow aeronauts. Yet rituals and spectacles, while emphasizing and praising the unique role of the aviators, used rhetoric and symbolism to anchor this uniqueness within the culture and history of the larger army. The values these rituals emphasized and promoted—selflessness, sacrifice, devotion—were enduring and timeless. The *voennye letchiki* thus symbolized all that was best of the Imperial military; they were merely a new incarnation of a uniquely Russian martial spirit. Their specific doctrine and duties might be new, but they were the progeny of a long and glorious tradition. As rituals are wont to do, aviation spectacles and funerals appealed to the timeless, invoked the mythic, and provided a sharp contrast to the rather inglorious present and downright ignominious recent past of the Russian army.

This was in part because the audience for such rituals was as much the officer corps and the army as a whole as it was the aviators themselves. Aviators in general did not need to be reminded of the dangerous nature of their work, nor of the courage and initiative necessary to lead even a dilettante seeking glory, rather than a sober minded professional, to take to the air. Likewise, the military pilots were active participants in their own rituals of both celebration and mourning, and no doubt profoundly felt more elation and loss than could be expressed in speeches or telegrams. In contrast, a large swath of the officer corps was mired in a peacetime malaise marked by stultifying routine, poor morale, and anxious self-examination. To them the aviation rituals presented a shining symbol. The *voennye letchiki* uniquely *were* at war, a war with the air, and their example was used to demonstrate that selflessness, sacrifice, and the ability to rise to a momentous challenge still existed in the heart of the Russian warrior. Aviation rituals thus implicitly distinguished between the sacred and the profane: the vistas of battlefields in the sky and the provincial muck of a regimental encampment, victory in the air and defeat and revolution on the ground.

The aviation spectacles and rituals were not only didactic, but often propagandistic. Aviation officers and commanders were acutely aware of the power of spectacle afforded by the airplane and the ways in which it could be used to legitimize their emergence to the army at large. Again, while the eloquent speeches were generally contributed by commanders, priests, and royalty, the line aviators themselves provided their own flourishes and contributions, performing flights and offering demonstrations for dignitaries and crowds. For unreconstructed senior officers of the line, it was a common complaint (though less so as aviation progressed) that the airplane was *only* good for spectacles and exhibitions, not serious military work; the numerous deaths involved only demonstrated its manifest unreliability. Ceremonies and performances provided a response to such criticism. They did far more than simply prove that airplanes could indeed fly; they played upon a sense of awe and wonder, which helped pave the way for the practical participation of aviation squadrons in maneuvers and tactical experimentation.

Rituals and spectacles functioned in a very similar way in regard to another audience: the public. Rhetoric and symbols worked to emphasize the community between the Imperial military and society. The values emphasized included selflessness, nobility, a deeper spirituality; the dominant theme was that Russians, as embodied in their pilots, already possessed skill, dedication, and knowledge equal to or better than the West. Moreover, such rituals assuaged duel concerns on the part of the military. On the one hand was the perceived gulf of understanding and trust between army and civilians, fruit of the past decade; on the other, throughout Europe the progress of aviation was accompanied by persistent fears of an ever more jaded aviation public, one increasingly immune to astonishment. Thus, spectacles and even funeral rituals served to galvanize the public's continuing commitment to the support of Imperial military aviation. Moreover, they did so in a way that precluded any troubling and divisive questions as to what form that support should take. Much like the elegant simplicity and powerful impression made by the flights of the aviators, a throng of enraptured or mourning citizens was a powerful yet uncomplicated display of empathy and unity.

The aviation rituals sought to emphasize the bond with yet another community: the regime itself. If the ideals exemplified by the *voennye letchiki* were traditional and timeless, so too was the object of that service. It is impossible to ignore the Imperial paeans that ran throughout rituals of both celebration and mourning. Col. Odintsov appealed to his patron, the Grand Duke, with praise and thanks, while priests and generals cited the solicitude of the Sovereign Emperor, who himself deigned to thank the aviators, in person and in writing, for their patriotic efforts. With an appreciative Tsar and a Grand Duke personally and actively involved in the challenge of developing military aviation, these rituals symbolically confirmed an image of a regime willing and seemingly able to master the challenges of modernization represented by the airplane.

Lastly, it is equally noteworthy what these rituals chose not to emphasize. Unlike the bellicosity of much of the civilian press when it expounded upon the need to develop aviation, Russian military aviation rituals smacked more of romanticism than social Darwinism. The enemy in this war was identified as the air itself; little mention was made of terrestrial enemies or their air forces. Likewise, the dead officers were canonized more as knights-errant on a noble quest, filled with all that was best and pure in the Russian soldier, rather than as warriors seeking to bring death to those beyond the Empire's borders. In part this merely reflected the infant status of military aviation. In part, however, such emphasis demonstrated that aviators spoke to concerns unique to the Russian officer community as a whole by providing a unifying example of virtue, competence, and confidence. Divisions aside, officers from all branches of arms could agree upon the symbols provided by the aviators, symbols which were a striking mix of the modern and the archaic, of advanced technology and timeless martial values.

A noted general from another time and place might well have reminded the aviators that one does not win a war by dying for one's country, but by making the enemy die for one's country. Nevertheless, this broader example was perhaps the most important contribution of military aviation rituals, and it mirrored the legacy of aviation to the army on a broader scale. Too small to directly touch more than a handful of officers, too new and untested to effect large scale tactical and strategic revolutions, the air forces and its aviators offered a powerful example of service and a shining symbol of redemption and hope. As the army continued in its efforts to gird itself for the test which many officers saw just over the horizon, almost all agreed that the aviators offered an inspiring and ideal symbol of the true mind, spirit, and soul of the Russian officer, and the ability of that officer to find Russian solutions to the challenge of modern warfare.

| Conclusion: Aviation, Military Culture, and Modernization

As the calendar turned to 1914, Russian military aviation, like the Imperial army itself, was at a crossroads. As they had for the past several years, enthusiasts both within the army and without trumpeted Russia's standing as second among aviation powers, behind only the universally acknowledged leader, France. The army, the state, and the nation as a whole had apparently heeded the urgent appeals to press forward in the building of a mighty air force. On paper Imperial Russia's aeronautical power was indeed impressive. By the outbreak of war in August 1914 the army disposed of 244 aircraft of various types deployed in 39 squadrons, along with 12 dirigibles. Some 300 military pilots had been trained since 1910, including 133 who were trained at Gatchina and Sevastopol and received the title of *voennyi letchik* in the first half of 1914.[1]

Although not all of the ambitious measures included in the official plan for the development of aeronautics and aviation had yet been implemented, further expansion was on the way. The remaining provisions of the plan were part of the Large Program of June 1914, the tsarist army's most ambitious attempt to answer the international challenges of technological adaptation and military modernization. The Program thus provided for the creation of 28 additional aviation squadrons and the formation of two new military aviation schools.[2] In light of the massive aircraft orders planned for 1914–1917, as well as the progress of domestic industry, "to contemporaries, the air services appeared to be entering a new era."[3] Perhaps of even greater portent, the Russian Empire had also produced heroes of world recognition and renown, aviators whose feats and innovations rivaled or surpassed those of other Europeans. Men like Mikhail Efimov and Staff-Capt. Petr Nesterov were proof positive of the Empire's reservoirs of talent, energy, initiative, and courage. The airplane indeed seemed to represent, in the words of General Kovan'ko, "the power, strength, and future of Russia."[4]

Yet, in many ways Imperial Russia's aeronautical might was merely a facade of modernity, a Potemkin village in the sky. A heterogeneous and aging complement of military aircraft was cause for concern, while various army reports complained of the uneven quality and lack of uniformity of parts regarding the aircraft and equipment produced in Russian factories. In a broader sense, the relative industrial weakness, poor infrastructure, financial poverty, and sheer expanse of the Russian empire continued to hinder the army in its efforts to supply its air forces and successfully exploit their potential. Likewise, the development of civilian aviation continued to lag, largely depriving the army of the ancillary resources available to its foreign allies and enemies. Thus, the same structural obstacles to its development that greeted the airplane in 1908 still largely remained six years later. Such difficulties were not resolved in the few months before the outbreak of war.

Further, the military aviation services themselves continued to be hampered by internal divisions and delays. Despite its efforts, the ambitions of the Aeronautics Unit of the General Staff remained largely unrealized. Although the its principal achievement, the comprehensive plan for the organization of the aeronautical and aviation services, was finally passed as part of the Large Program, the outbreak of war rendered it moot. At the same time, instead of General Shishkevich's proposed reorganization and expansion of the Aeronautics Unit, it was itself liquidated in early 1914. Daily oversight of the aviation and aeronautical services and their technical needs was transferred to a new Aeronautics Section of the newly reconstituted Main Military Technological Administration [GVTU], successor to the Main Engineering Administration. The General Staff retained overall organizational control and strategic direction of the aviation squadrons.[5]

An explanation of this reorganization in *Razvedchik* implicitly recognized the contradictions that marked the brief history of the Aeronautics Unit. It noted that technical and logistical questions fell more naturally within the competence of the Military Technological Administration, as GUGSh possessed neither the personnel nor resources to adequately address such questions.[6] In short, the attempt of the Aeronautics Unit to decide simultaneously broad questions regarding the strategic use and overall organization of the aviation and aeronautical services, as well as more narrow questions regarding equipment, proved beyond its professional purview. The article neglected to mention deeper problems. Despite the creation of the Aeronautics Unit, both the Department of the Air Force and the Officers' Aeronautics School were largely able to define their own goals and priorities and retain an active and somewhat independent role in the realm of training pilots and testing new technology. Thus, the liquidation of the Aeronautics Unit was a recognition that it had largely failed in its mission to achieve unity and coordination in the development of military aviation. The General Staff instead reverted to the function that some argued it should have assumed in the first place, namely identifying the strategic and tactical role of aircraft in future war, while leaving the specifics of equipment, personnel, and administration to experts in the field.[7] It was therefore the new Aeronautics Department of the GVTU that would oversee aviation in the scant few months until the outbreak of war in August 1914, at which point the mounting demands of mobilization soon spawned a variety of expedients to meet the army's wartime aviation needs.

On the eve of war, both persuasive progress and lingering concerns bore witness to the hopes and fears which attended the birth of tsarist military aviation. In terms of its demands and potential, the airplane clearly represented far more than a mere advance in weapons technology, and hence necessitated significant adaptations in the realm of finance, production, supply, and administration. Much like early railroad development, realizing the military potential necessarily involved the dedicated efforts of state, society, and industry. At the same time, the rapidity and cosmopolitan character of international aviation development placed limits on any purely independent response to the airplane. Regardless of their provenance, theoretical discoveries, technological advances, innovations in piloting, and organizational adaptations were quickly studied and frequently copied by all of the European military air arms. They too faced the challenges of defining and utilizing the role and power of both airplane and aviator within the charged context of military competition in the decade before World War I.

In terms of its rapidity, scope, and intensity, the development of aviation thus proved uniquely emblematic of the challenges of modernization for tsarist military servitors, the regime they served, and the Russian Empire as whole. The airplane was hailed by contemporaries as the quintessential symbol of modernity, the apotheosis of humanity's efforts to subjugate nature. In turn, the response to the airplane within the Russian Imperial army is particularly revealing not merely because of the largely military orientation of pre-World War I aviation, but because of the unique nature of its officer community. Like their brethren throughout Europe, Russian officers possessed an elaborate and rigidly enforced social system. What is frequently implicit or unspoken in civilian social relations—hierarchy, deference, role, function, identity—was made explicit in rank, dress, regulations, codes of conduct, and ritual. Thus, within the institutional, social, and cultural confines of the tsarist military community, values, traditions, and particular canons of service framed responses to aviation for both supporters and foes alike. Those officers who were conscious of the airplane's significance endeavored to exploit it in a way that respected the customs, traditions, and virtues of the Russian Imperial army. Yet, defining the essence of those traditions and articulating the limits of that respect produced intense debate. The disparate responses to the challenge of aviation reflected in concentrated form the tensions involved in implementing reform and defining modernization in the post-1906 officer corps as a whole. They provide insight into the contentious currents of thought regarding tsarist officers' sense of identity, role, duty, and service, as well as the increasingly complex relationship of the officer community to the broader social and cultural milieu of late Imperial Russia.

The unique nature of the Russian Imperial response to military aviation can be seen most clearly in comparative context. As occurs with most new technologies, all of the European militaries experienced their own internal disagreements and divisions in the realm of aviation. In the French army engineers and artillerymen adhered to different doctrines regarding the deployment of airplanes; in Britain, aviation development saw a division of effort and organization between army and navy; in Germany a lingering predilection for dirigibles initially retarded aviation technology.[8] Yet, despite such divisions, scholars have argued that in the broader realm of community and identity, the airplane was assimilated more or less seamlessly into the prevailing military cultures of the European powers. The ethos of revanche and offensive *a outrance* that dominated the French army, the liberal militarism and commitment to the economical application of force that characterized the British military establishment, and the commitment to the professional exploitation of the most modern technology among the German General Staff all proved uniquely receptive to the promise of aviation. At the same time, all of these armies, to varying degrees, found common cause with important segments of civilian society in the development of military aviation. Soviet historiography and memoirs, as well as Western studies, largely present a very different picture in regard to Russian military aviation and the role of tsarist officer culture. They cite a handful of energetic, progressive officers who were convinced of the airplane's primacy and recognized that it generated a whole complex of new demands, priorities, and models of service. However, despite the efforts of such officers, these broader demands were fundamentally irreconcilable with the prevailing culture of backwardness and inertia within the army writ large.

Yet the relationship between the culture of the Russian officer community and the airplane was actually far more complex and dynamic, marked not by any simple ascriptions of cultural resistance or reflexive imitation. Despite being informed by the cosmopolitan context of aviation development, the responses of tsarist officers to the airplane were necessarily framed by the unique social, economic, and geographical context of the Russian Empire itself. As astute observers recognized, factors such as the absence of capital, poorly developed infrastructure, and a meager industrial base constrained options and presented difficult choices, placing real limits on any simple appropriation of foreign models. In turn, such concerns were couched within an older and broader discourse about military "modernization" in the Russian Empire. The principal themes again revolved around juxtapositions of Russia and the West, imitation and contamination, the imperatives of progress and the legacies of backwardness, themes that took on a unique caste in the realm of aviation. Yet the airplane to an extent rendered irrelevant familiar tensions between foreign models and Russian uniqueness. While aviation was a profoundly international development, from its inception Russian soldiers and civilians played an active and at times leading role in its development. The airplane was indeed as much a Russian technology as a European or American one; hence, tsarist servitors needed to bow and scrape to no one, nor rationalize borrowing or backwardness. Discussion about the merits and course of military aviation was thus free from much of the wider acrimony attending other debates about technological appropriation or foreign imitation.

The numerous social, cultural, and philosophical divisions within the Russian military community itself also mediated the response to the airplane. While a lack of monolithic agreement regarding mission and doctrine prevented any universal embrace of aviation, this further ensured that there was no systematic rejection of aviation. In a time of discord and debate, it was difficult to define the nature of heresy. The airplane did not threaten to overturn cherished military traditions or undermine a uniquely Russian military art; in its infancy the aviation service was not a danger to vested institutional interests. The constituency of engineers and especially aeronauts was itself small and largely marginal, while perhaps the most preeminent institution within the army, the General Staff, ultimately appropriated jurisdiction over aviation development. Moreover, the nature of aviation itself served to militate against possible resistance in the Russian officer community. The union of pilot and airplane tended to challenge traditional juxtapositions of technology and morale. Marked by voluntarism and initiative, the role of pilot fit well within a military culture that had traditionally valued personal accomplishment and was still deeply suspicious of rigid frameworks of doctrine and dogma.

The unsettled culture of the post-1906 tsarist officer community served not just to minimize resistance to the airplane, but also to provide channels of active support. In a negative sense, slow rates of promotion, a regimental economy dominated by bureaucratic routine, and the resulting ineffectual character of line service frequently provided the impetus for junior officers to volunteer for aviation service. In a more positive sense, the tsarist army offered supporters of the airplane broad space and opportunity in which to pursue their ambitions. The nascent air force was largely unencumbered by an institutional history and tradition; aviation thus offered a new arena of initiative and activity for the expert, the enthusiast, and the initiate alike. In the context of humiliation that followed the events of 1904–06, the redemptive promise of future power represented by the airplane took on heightened appeal. In like fashion, within the broader context of the unique difficulties of geography, infrastructure, and economy within the Russian Empire, the airplane offered greater rewards to Russia than perhaps any other power.

The relationship between tsarist officer culture and the airplane was thus dynamic and reciprocal. If the challenge of aviation was interpreted within the unique social and cultural context of the officer community, this community itself was to an extent transformed by the experience. In particular, the power of symbolism that characterized aviation discourse was seized upon by varied groups to advance their own distinct agendas. Some officers, convinced of the future primacy of the airplane for both army and state, advanced urgent arguments for its rapid and comprehensive development. Others used aviation as a forum for broader critiques of prevailing structures of authority and duty within the army. The result was a host of competing appraisals and projects regarding the proper utilization of the airplane.

While they shared the same ultimate goals of aerial power and greatness, the competing visions of aviation development were quite different in their assumptions and approaches. They were largely embodied in institutional adaptations unique to the Russian army: a tripartite division between a military school of engineering parentage, a specially created organ of the General Staff, and a flying school administered under the auspices of a quasi-public organization and an imperial Grand Duke. In no other European army did the General Staff or an imperial scion take so active and direct a role in the development of aviation. There was far more at stake in this than matters of administrative efficiency or questions of policy. Instead, these different institutions were emblematic of competing cultures of duty and authority, manifested in divergent priorities, philosophies, and goals. These institutions and their leading officers thus contested the terms upon which an ethos and culture of aviation service would be constructed. The available paradigms included the distinct strains of professionalism represented by the Engineering Department and the General Staff, the patrimonial authority embodied in the Grand Duke and the Department of the Air Force, and, increasingly, a new and unique ideal of aviation professionalism.

This last paradigm was potentially most disruptive. Harley Balzer locates the issue of professionalism at the heart of the familiar conundrum that "from the time of Peter the Great, Russian rulers wanted formal and technological modernization without the accompanying social consequences."[9] Such attempts to tightly manage modernization were invariably unsuccessful. The tsars and their ministers instead were rewarded with more or less assertive of groups of professionals who advocated their own agendas and institutional interests. To be sure, military professionals had different concerns than civilians. While William Fuller notes that military lawyers and engineers had corresponding magnets of civilian professionalism in their fields, they remained Imperial officers. Service to the state was their creed and raison d'être; breaking free of its control and tutelage was to an extent unthinkable.[10] Yet, even within the relatively "safe" and loyal confines of the higher reaches of the army, there was a latent fear of "alien" concepts of military professionalism and its implied autonomy.

This peril was manifested most vividly in the process whereby for many such officers professionalism increasingly implied politics. If the professions in late Imperial Russia defined their service to an abstract "public" and common good, then such officers' conception of service increasingly stood for the abstractions of "state" and "nation." Notably, Fuller observes that although War Minister Rediger's plan for a national military reform had been defeated with his replacement by Sukhomlinov, the principles it espoused survived: a fully professional army, "consecrated to defending the state against external enemies alone, while concomitantly striving to win the affection of the Empire's population." These principles, in tune with Guchkov's and the Octobrists' "unabashed imperialism, pride in Russia's status as a Great Power, and a desire for a large and strong army" continued to be very attractive to elite military professionals within the army.[11]

Such concerns particularly resonated with the senior officers of the aviation service. The projects they pursued and the methods they employed illustrated the influence of conflicting conceptions of competence and authority, revealing the tensions as well as the linkages between professional conceptions of duty and the enduring appeal of patrimonial ideals of military service, framed within a culture of hierarchy and deference. Professional officers of the aviation services faced numerous obstacles in asserting their ideals of service: a reluctance to delegate neutral spheres to express dissent, the habit of equating such dissent with insubordination, the impetus to deride initiative as base careerism. Within this atmosphere, the result was an uneasy mix of appeals to education, training, and objective standards of expertise, as well as the exercise of prerogatives which inhered in rank, individual personality, and honor. When such officers' ambitions were not met, the objects of their frustration were resistant aeronauts, apathetic commanders, or "idle *chinovniki,*" to use General Kovan'ko's term.[12] Even professionally informed conceptions of service might manifest themselves in rivalry, intrigue, and the defense of personal prerogative. As Fuller emphasizes, professionalism should be measured as much in terms of consciousness than "objective" results. Thus, negative corporatism, professionalism, and patrimonialism could coexist within the minds of officers. Tradition and novelty, custom and innovation, expertise and honor, were not irreconcilable.

The tension between competing ideals of duty, service, and identity was also clearly illustrated at the level of the line within the aviation service. The contrasts between contending paradigms of service were to an extent embodied geographically and culturally in the two competing centers of aviation training, the Aviation Section of the Officers' Aeronautics School and the Sevastopol Aviation School of the Department of the Air Force. The first school was notably professional, appealing to objective standards of science and knowledge; the other school invoked more traditional forms of officer identity, aided by the royal imprimatur of a Grand Duke. One was primarily concerned with issues of competence and expertise, while the other emphasized an urgent sense of noble mission and traditional glory. Yet the motives and goals of individual volunteers were not necessarily reducible to either of these contending paradigms. The decision to volunteer for aviation service could either attest to the underlying vitality of officer culture or the failure of the army community to generate meaningful service in established branches of arms. Aviation promised reknown, glory, or merely an opportunity for meaningful service for those seeking to break out of stultifying routine or the shackles of negative corporatism. Such considerations unique to the opportunities of flying were also frequently infused with broader concerns of class, honor, and patronage. The distaste for the crassness of European culture, manifested in commercial impulses, was thus a powerful unifying factor for military aviators. Likewise, when officers differentiated themselves from civilians, even their fellow civilian aviators, they appealed to notions of selflessness, duty, and sacrifice shared by their comrades in arms.

Tsarist officer culture thus provided both centrifugal and centripetal forces within the community of military aviators. Drawn to aviation service from various branches, units, and sub-cultures of the army, some officers sought fate, some sought glory, while some sought to advance the technology and theory of flight. The Imperial air force attracted numerous committed and energetic volunteers and turned out competent, devoted, innovative aviators, some with international reknown. It also turned out others who would not fly one minute beyond what was necessary to earn their flight pay. Russia's new corps of aviators was distinguished by different training experiences and philosophies, divided by different motivations and degrees of devotion to their cause, and riven by broader factors of class, nationality, and honor which informed life in the officer corps as a whole.

Nevertheless, the Russian aviation service was ultimately marked by a vibrant and ascendant, if contested, commitment to professionalism, construed in terms of strict admissions, rigorous training, commitment to constant inquiry, and an awareness of the unique demands of military aviation. Soviet historiography sought to emphasize the numerous obstacles faced by such professional officers, represented in the Scylla and Charybdis of fashionable playboys and obsolete aeronauts, as well as the institutionally entrenched backwardness of the army leadership. However, this study demonstrates that it was indeed professional military aviators like Staff-Capt. Nesterov who largely determined the standards of what it meant to be a *voennyi letchik.* Nesterov and his like-minded comrades dominated the discourse of aviation in reports and in the press; their opponents provided a largely negative model of passive resistance. Moreover, heroes like Nesterov did not deny worthy traditions nor denigrate cherished values. Instead, they appeared to combine all that was best, timeless and modern, in the Russian Imperial officer. It is also difficult to see these examples of aviation professionalism as anomalies within the Imperial army, while deriding the flyboys and *beloruchki* as typical products of socialization. The career profiles, educational experiences, and personal backgrounds of the officers cited throughout this study make such simplifications untenable. They were all products, in varying fashion, of the Imperial military community.

Further, it must be emphasized that those who resisted the professional ideals of a Nesterov or Efimov were not necessarily retrograde. In time when much of the theory and practice of aviation was still novel and obscure, those who believed in aviation as art, who preached innate talent and caste-informed notions of identity, and who argued for strict functional specialization as well as rigid delineation between the roles of pilot and mechanic, officer and soldier, had their analogs in the air forces of the other European powers. Though not professionals in Fuller's sense of the term, this does not mean such officers were necessarily devoid of devotion to duty. Dilettantes and poseurs there were. Yet alongside them were serious pilots with very different conceptions of duty and honor, conceptions which militated against the adoption of the strict professional paradigm of aviation service exemplified by Nesterov.

These various strains of tsarist aviation culture demonstrate that there were indeed reservoirs of commitment, competence, and purpose even among demoralized officers of the line. In light of much of the historiography of the late Imperial army, this point bears emphasizing. Moreover, as with the larger debate regarding technology and culture within the Russian officer community, distinctions between the professional and traditional paradigms of duty and service should not be drawn too sharply. Very few officers would deny outright the importance of training and education, while thoroughly professional officers would not deny the continuing value of honor, sacrifice, and loyalty to the Emperor. It would take the dynamics of World War I and the regime's prosecution of the war effort to make such juxtapositions stark and largely irreconcilable. Until then, officers like Nesterov and the Cossack Tkachev were lauded precisely because they embodied the harmonious integration of modern technology and Russian martial traditions.

Such multi-faceted responses to the airplane were replicated in the broader forum of the army as a whole. Military aviation attracted numerous supporters and patrons who saw in the airplane the dawning of a new epoch in military power. Some of these officers took an active part in efforts to realize the airplane's transformative potential; others merely argued for the guardians of lethargy and routine to allow an unfettered "path to the future" for the young, uncorrupted aviators. Still other vested constituencies welcomed the advent of the airplane but sought to arrogate to themselves the overall right to manage and utilize its power. Most notable here were officers of the General Staff who sought to impose their own ideal of leadership upon the aviation service, one informed by a well-developed culture of duty, mission, and professionalism. While aviators indeed had to struggle against perceived shortsightedness, inertia, even ignorance, on the part of both their immediate and senior commanders, the widespread resistance to aviation on the part of the military leadership, so important to Soviet historiography, is difficult to locate in the sources. Instead, Russian aviatiors seemed to find roughly the same proportions of supporters and skeptics as their brethren throughout European armies. Moreover, one could find some justification for those officers who questioned aviation, both regarding the airplane's potential application as a weapon and in broader terms of its demands on attention, resources, and effort. They should not summarily be dismissed, as is often the case, as mere obscurantists. If many of those resistant to aviation had no firsthand knowledge of the theory and technology of flight, perhaps they did have a firsthand knowledge of the enormous problems of strategy, organization, staffing, and supply that occupied the Russian army in these years.

It was precisely here that ritual was so important. Ceremonies and spectacles of both celebration and mourning functioned to deny such dissonance by emphasizing the unity and harmony of all in a great patriotic cause. Aviation rituals lauded the efforts of this brave new cadre of military aviators while placing their example firmly within the proud history, traditions, and culture of the Russian Imperial army. As such, they offered a powerful, integrative, and unifying vision. Timeless Russian martial ideals of sacrifice, individual genius, and the will to conquer were united with the demands of modern technology. The airplane thus promised to negate divisions between different branches of arms, between technology and morale, between tradition and imitation, between backwardness and progress. The airplane indeed provided a powerful symbol of redemption and renewal, even if the very elasticity of this symbolism dulled its potential to destabilize traditional conceptions of duty and identity. It was precisely because aviation seemed to transcend stark oppositions of tradition and modernization, of uniqueness and progress, that the need to question prevailing paradigms of duty and authority was lessened. Indeed, examination of the aviation rituals suggests that perhaps the airplane's greatest influence and most important legacy for tsarist officers rested in the realm of culture. Such rituals also revealed the limits of that legacy. The *voennye letchiki* promised to uphold the best traditions of the Russian army while simultaneously demonstrating its ability to rise to the challenge of technological modernization and maintain the Empire's place amongst the powers of Europe. Yet the actual process of building a cadre of military aviators demonstrated how difficult and contentious it was to fill those ideals with any unified and coherent content.

Overall, therefore, the tsarist officer community responded quite well to the challenge posed by the airplane. It provided a unique social and cultural setting that functioned to increase the attraction and relevance of aviation and to provide opportunity for the exercise of initiative in this field. At the same time, divisions within this community clearly placed definite limits upon the fruits and productivity of this initiative. The brief history of the Imperial air force thus reveals no structural resistance to modernization, nor intractable backwardness, but illustrates instead the pains of modernization itself. The specific case of aviation thus fits well into the broader historical experience of the Russian Imperial army. Throughout the late nineteenth and early twentieth centuries, tsarist officer culture did not fatally interfere with military modernization, insofar as modernization is specifically defined as the acquisition of up-to-date military technology and its integration into existing military structures. The relative weight and culpability of officer culture as an impediment actually seems small in contrast to the constraints of finance, an industrial base that was simply not competitive with the West, and Russia's enormous and conflicting strategic responsibilities and burdens.[13]

With the Russian Imperial army's entry into World War I, broader structural weaknesses in the Imperial economy, industry, and infrastructure combined to produce a situation all but intractable as the war progressed. Despite a rapid expansion of the air force, its training schools and facilities, and its administrative institutions, Russian military aviation was unable to overcome material constraints and handicaps. The interruption of trade in technology and equipment with France, the inability of domestic industry to make up for all of the shortfall, large distances and resulting wear and tear on equipment, and the high rate of attrition among pilots ultimately carried far more weight than individual successes and exploits, of which there were many.[14] Thus, even as the aviation services rapidly expanded, Russian military aviation was itself in a downward spiral from the first day of war.

Given these broader constraints, the airplane's effect upon broader military structures, practices, and doctrines within the tsarist military was indeed limited. Again, its greatest legacy was perhaps symbolic, in that it offered a powerful example of the ability of the Russian army to master the challenges of modernization, but do so in a uniquely Russian, indeed an Imperial Russian, manner. This nascent culture of Russian military aviation would transcend the Imperial/Soviet divide. Its defining characteristics included an emphasis upon the cultivation of human cadres and resources in the face of material poverty; a flexibility in terms of individual experimentation and initiative; a desire for public assistance and cooperation, accompanied by a fear of the selfish and egotistical motivations that might accompany such support; a deep suspicion of both capitalist motivations and reliance upon foreigners; and thus a firm realization of the need to cultivate domestic resources and productive capabilities. This culture would be carried forth in the surviving institutions and personnel of the tsarist air services that helped form the nucleus of the Soviet Red air force.

Neither Staff-Capt. Petr Nesterov or Mikhail Efimov lived to witness the Soviet commitment to modernity in the skies. Convinced that it was theoretically possible to successfully ram and destroy an enemy aircraft, Nesterov attempted this maneuver during the first month of fighting in WWI. Angered by an Austrian airplane flying unmolested above the Russian lines, Nesterov took off and engaged the enemy. He indeed succeeded in ramming the Austrian craft, bringing his enemy down, but was himself killed when his own plane crashed as well. Mikhail Efimov would continue both to serve diligently and suffer various indignities at the hands of his aristocratic nemeses in the Russian air forces. Despite a varied and distinguished record of service in the war, he never received an officer's commission. His personal and professional experiences no doubt shaped his decision to join the Bolsheviks in the Civil War; caught behind enemy lines, he was captured and shot by White forces in the south of Russia in 1920.[15] Both men would later be claimed by the Soviets as martyrs to Russian aviation.

| Notes

Introduction

1. See L. V. Ruzov and Iu. N. Iablochkin, *Gatchina. Istoricheskii ocherk* (Leningrad: Lenizdat, 1959); *Aviatsiia v Rossii: [Sbornik]. K 100-letiiu otechestvennogo samoleto-stroeniia* (Moscow, 1983); *Kryl'ia rodiny: [Sbornik]* (Moscow, 1983); and L. M. Kuz'mina, *Pervye kry'lia Rossii: Stranitsy vekovoi istorii aviapredpriiatiia* (Moscow: Voennoe Izdatel'stvo, 1993).
2. A revised edition of this study was published in 1995, but the tone and primary arguments of the work remain largely unchanged. Petr D. Duz', *Istoriia vozdukhoplavaniia i aviatsii v Rossii: (Period do 1914 g.)* (Moscow: Mashinostroenie, 1995). A second volume devoted to the war years has also been revised: Duz', *Istoriia vozukhoplavaniia i aviatsii v Rossii: iiul' 1914 g.– oktiabr' 1917g.* (Moscow, Mashinostroenie, 1986).
3. See, for example, David R. Jones, "The Beginnings of Russian Air Power, 1907–22," in *Soviet Aviation and Airpower: A Historical View*, eds. Robin Higham and Jacob W. Kipp (Boulder, CO: Westview Press, 1977). More general works which address pre-war Russian aviation include James Stokesbury, *A Short History of Air Power* (New York: William Morrow and Co., 1986), and John H. Morrow, *The Great War in the Air: Military Aviation from 1909 to 1921* (Tuscaloosa, AL: University of Alabama Pres, 1993).
4. Robert A. Kilmarx, *A History of Soviet Air Power* (New York: Prager, 1962), 3–25.
5. Scott W. Palmer, *Dictatorship of the Air: Aviation Culture and the Fate of Modern Russia* (New York: Cambridge University Press, 2006).
6. Palmer, "On Wings of Courage: Public 'Air-Mindedness' and National Identity in Late Imperial Russia," *Russian Review* 54, 2 (April 1995): 209–26.
7. For more on the great Sikorsky, see K. N. Finne, *Igor Sikorsky, The Russian Years* (Washington, D.C.: The Smithsonian Institution Press, 1987) and Dorothy Cochrane, et al, *The Aviation Careers of Igor Sikorsky* (Seattle, WA: Washington University Press, 1989).
8. Von Hardesty, "Early Flight in Russia," in *Russian Aviation and Air Power in the Twentieth Century*, eds. Robin Higham, John Greenwood, and Von Hardesty (Portland, OR: Frank Cass, 1998), 18–36.
9. This burst of literature in the late 1980's was covered in several review articles. See Peter Kenez, "Russian Soldiers and Military Conflicts," *Slavic Review* 45 (Winter 1986): 720–723; Walter Pintner, "The Imperial Russian Army: Society, Politics, Economics, and Revolution," *Russian Review* 46 (October 1987): 439–46; and Michael Perrins, "Army Politics and Russian Society 1462–1937," *European History Quarterly* 17 (April 1987): 219–23.
10. See John L. H. Keep, *Soldiers of the Tsar: Army and Society in Russia, 1462–1874* (New York: Oxford University Press, 1985).

11. See Istvan Deak, *Beyond Nationalism: A Social and Political History of the* Hapsburg *Officer Corps* (New York: Oxford University Press 1990); Karl Demeter, *The German Officer-Corps in Society and State, 1650–1945*, trans. Angus Malcolm (New York: Weidenfeld and Nicoloson, 1965); and also Steven E. Clemente, *For King and Kaiser! The Making of the Prussian Army Officer, 1860–1914* (New York: Greenwood Press, 1992).
12. John Shy, "The Cultural Approach to the History of War," *Journal of Military History* 57, no. 5 (October 1993), 25.
13. Sociologically oriented models of civil-military relations and military professionalism are often informed by a conception of "militarism" as a dysfunctional civil-military relationship, dominated as it is by the latter; they usually thus concerned with state structure and high politics, not the daily experience of the officer corps as a whole. To cite just a few examples, see Samuel P. Huntington, *The Soldier and the State* (Cambridge: Harvard University Press, 1957; Morris Janowitz, *The Professional Soldier, A Social and Political Portrait* (Glencoe, IL: Free Press, 1960); G. Teitler, *The Genesis of the Professional Officers' Corps* (Beverly Hills: Sage, 1977); Jacques van Doorn, *The Soldier and Social Change: Comparative Studies in the History and Sociology of the Military* (Beverly Hills: Sage, 1975); and Samuel E. Finer, *The Man on Horseback: The Role of the Military in Politics* 2nd ed., (London: Pall Mall Press, 1962). See also Edward Coffman, "The Long Shadow of The Soldier and the State," *Journal of Military History* 55 (January 1991): 69–82.
14. Alfred Vagts, *A History of Militarism: Romance and Realities of a Profession* (New York: W.W. Norton, 1937), 11.
15. Fuller, *Civil-Military Conflict*, 5–6.

Chapter 1

1. Dates for key international events are cited according to the Gregorian calendar. Archival and secondary Russian documents from this period are dated in the Julian calendar.
2. For an excellent recent collection concerning the war, see John W. Steinberg, et al., *The Russo-Japanese War in Global Perspective: World War Zero* (Boston: Brill, 2005).
3. Alan Wildman, *The End of the Russian Imperial Army: The Old Army and the Soldiers' Revolt (March–April 1917)* (Princeton: Princeton University Press), 58-60.
4. See E. Willis Brooks, "Reform in the Russian Army, 1856-1861," *Slavic Review* 43 (Spring 1984): 63–82.
5. The traditional analytical model of military reform as linear cause and effect is itself problematic and its assumptions debatable. Armies are notorious for being fundamentally conservative institutions. As E. Willis Brooks' article demonstrates, even a defeat perceived to be as dramatic and traumatic as the Crimean War did not in itself result in immediate sweeping reforms in the Russian army. See Brooks, "Reform in the Russian Army." Comparative evidence further demonstrates the problems associated with the traditional knee-jerk model of reform as a response and reaction to defeat or new technology. See, for example, Richard Stoll, "Steaming in the Dark? Rules, Rivals, and the British Navy, 1860–1913," *Journal of Conflict Resolution* 36 (June 1992): 263–83, for one amusing take on the impact of new technology on a military organization.
6. See Bruce Menning, *Bayonets Before Bullets* (Bloomington: Indiana University Press), chapters 6-7; William C. Fuller, *Civil-Military Conflict in Imperial Russia 1881–1914*

(Princeton: Princeton University Press, 1985), chapter 7; Wildman, *The End of the Russian Imperial Army*, chapters 1–2; and A. Kersnovskii, *Istoriia russkoi armii* (Moscow: Golos, 1992).

7. See Menning, *Bayonets Before Bullets*; Fuller, *Civil-Military Conflict*; Fuller, *Strategy and Power in Russia*; Walter Pintner, "The Imperial Russian Army: Society, Politics, Economics, and Revolution," *Russian Review* 46 (October 1987): 439–46, and Pintner, "Russian Military Thought: The Western Model and the Shadow of Suvorov," in *Makers of Modern Strategy from Machiavelli to the Nuclear Age*, ed. Peter Paret (Princeton: Princeton University Press, 1986), 354–75; Carl Van Dyke, *Russian Imperial Military Doctrine and Education 1832–1914* (New York: Greenwood Press, 1990); Jack Snyder, *The Ideology of the Offensive: Military Decision Making and the Disasters of 1914* (Ithaca: Cornell University Press, 1984); Walter T. Wilfong, "Rebuilding the Russian Army, 1905–1914. The Question of a Comprehensive Plan for National Defense" (Ph.D. diss., Indiana University, 1977); Peter Von Wahlde, "Military Thought in Imperial Russia" (Ph.D. diss., Indiana University, 1966).
8. Walter Pintner, "The Burden of Defense in Imperial Russia, 1725–1914" *Russian Review* 43 (July 1984): 251; for a broader analysis of such issues, see William Fuller, *Strategy and Power in Russia , 1600–1914* (New York: Free Press, 1992).
9. Menning, *Bayonets Before Bullets*, 210. See also Pintner, "Russian Military Thought."
10. John Bushnell, *Mutiny Amid Repression* (Bloomington: Indiana University Press, 1985), 1–2.
11. V. A. Zolotarov, "Russian Military Thought at the Turn of the Twentieth Century," *War and Society* 9:1 (May 1991), 58–9. See also Pintner, "Russian Military Thought," 356–57.
12. Peter Kenez, "A Profile of the Prerevolutionary Officer Corps," *California Slavic Studies* 7 (1973): 152. For more on Dragomirov, see Menning, *Bayonets Before Bullets*, chapters 1 and 4, and von Wahlde, "Military Thought in Imperial Russia."
13. Menning, *Bayonets Before Bullets*, 200–37; Van Dyke, *Russian Imperial Military Doctrine*, 132–54.
14. Menning, *Bayonets Before Bullets*, 204.
15. Kenez, "A Profile of the Prerevolutionary Officer Corps," 153.
16. See Walter Pintner, "The Burden of Defense in Imperial Russia, 1725–1914."
17. See Menning, *Bayonets Before Bullets*, chapter 7, for a good review of such measures.
18. Menning, *Bayonets Before Bullets*, 219–21; Fuller, *Civil-Military Conflict*, 222–25.
19. Again, see Menning, *Bayonets Before Bullets*, chapter 7.
20. Pintner, "Russian Military Thought," 375.
21. Menning, *Bayonets Before Bullets*, 207–08.
22. Capt. Legara, "Nravstvennaia podgotovka voisk v mirnoe vremia," *Voennyi sbornik* no. 4 (April 1911): 87–92.
23. See again the military journals *Voennyi Sbornik* and *Razvedchik* for a fuller discussion.
24. See Wildman, *The End of the Russian Imperial Army*, 38.
25. Kenez, "A Profile of the Prerevolutionary Officer Corps," 123–28; Wildman, *The End of the Russian Imperial Army*, 14–16, 20–22.
26. William Fuller, *Civil-Military Conflict in Imperial Russia,* 16.
27. Ibid., 16. Fuller notes that almost all memoirs of Russian military men distinctly emphasize this point.
28. Ibid., 22–29.

29. Kenez, "A Profile of the Prerevolutionary Officer Corps," 131.
30. Ibid., 131.
31. While the pre-war officer corps is not really his focus, Allan Wildman's opening chapter of his magisterial *The End of the Russian Imperial Army* offers a fine summary of such views.
32. Aleksandr Kuprin, *Sobranie sochinenii* (Moscow: Pravda, 1964).
33. See Gen. A. I. Denikin, *The Career of a Tsarist Officer: Memoirs 1872–1916*, trans. Margaret Pototski (Minneapolis: Universitry of Minnesota Press, 1975), and Denikin, *Staraia armiia*, 2 vols. (Paris: Rodnik, 1929–1931).
34. John Bushnell, "The Tsarist Officer Corps, 1881–1914: Customs, Duties, Inefficiency," *American Historical Review* 86, no. 4 (October 1981): 753–80.
35. Ibid., 753, 773.
36. See Menning, *Bayonets Before Bullets*, 215–16; Kenez, "A Profile of the Prerevolutionary Officer Corps," 152–53; Bushnell, "The Tsarist Officer Corps," 775–77; and Denikin, *Staraia armiia.*
37. For example, Bushnell notes that while coming from a far more sympathetic source, the novel by the emigre general Petr Krasnov, *Ot Dvuglavogo Orla k krasnomu znameni* (Berlin: Gutnov, 1921), describes customs and behavior among Tsarist officers remarkably similar to those described by Kuprin, despite the fact that the units in question were so different. Ibid., 754.
38. Bushnell, "The Tsarist Officer Corps," 762.
39. Ibid., 760–62.
40. Fuller, *Civil-Military Conflict*, 26–29.
41. Kenez, "A Profile of the Prerevolutionary Officer Corps," 140, 136. An invaluable source on this topic, Kenez also breaks down the social and religious composition of the officer corps in the last decades of the empire. Kenez, "A Profile," 132–33, 137–40.
42. Wildman, *The End of the Russian Imperial Army*, 22-23. Wildman also provides a table of social composition comparing the years 1895 and 1911 which generally confirms Kenez's figures.
43. Kenez, "A Profile of the Prerevolutionary Officer Corps, 140–41, 157–58.
44. Ibid., 141–42.
45. Wildman, *The End of the Russian Imperial Army*, 22.
46. Kenez, "A Profile of the Prerevolutionary Officer Corps," 143.
47. Wildman, *The End of the Russian Imperial Army*, 24, 72.
48. Kenez, "A Profile of the Prerevolutionary Officer Corps," 133–34.
49. Bushnell, "The Tsarist Officer Corps," 765.
50. Fuller, *Civil-Military Conflict*, 32.
51. Fuller, *Civil-Military Conflict*; Matitiahu Mayzel, "The Formation of the Russian General Staff, 1880–1917: A Social Study," *Cahiers du Monde russe et sovietique* 16, nos. 3–4 (July/Dec. 1975): 297–320; Mayzel, "Generals and Revolutionaries, the Russian General Staff during the Revolution: A Study in the Transformation of a Military Elite," in *Studien zur Militargeschichte, Militarwissenschaft und Konfliktforschung*, (Osnabruck) 19 (1979). On the General Staff, see also John Steinberg, "Russian General Staff Training and the Approach of War," in *Authority, Identity, and the Social History of the Great War*, ed. Frans Coetzee and Marilyn Shevin-Coetzee (Providence, RI: Berghahn Books), and Steinberg, *All the Tsar's Men: Russia's General Staff and the Fate of the Empire, 1989–1914* (Baltimore: Johns Hopkins University Press, 2010).

52. Fuller, *Civil-Military Conflict*, 30.
53. Ibid., 30.
54. Bushnell, "The Tsarist Officer Corps," 774.
55. Fuller, *Civil-Military Conflict*, 30.
56. Kenez, "A Profile of the Prerevolutionary Officer Corps," 141.
57. See for example Josh Sanborn, *Drafting the Russian Nation: Military Conscription, Total War, and Mass Politics, 1905–1925* (DeKalb, IL: Northern Illinois University Press, 2003); David Schimmelpenick van der Oye and Bruce Menning, eds., *Reforming the Tsar's Army: Military Innovation in Imperial Russia from Peter the Great to the Revolution* (Washington, D.C.: Woodrow Wilson Center Press, 2004); David Rich, *The Tsar's Colonels: Professionalism, Strategy, and Subversion in Late Imperial Russia* (Cambridge, MA: Harvard University Press, 1999).
58. Denis Showalter, "Imperial Russia and Military History," in *Reforming the Tsar's Army*, 326–27.
59. Fuller, *Civil-Military Conflict*, 46.
60. A. Dmitrevskii, "Vliianie akademicheskogo obrazovanie na nravstvennyi element (silu dukha)," *Voennyi sbornik* no. 5 (May 1914), 61–62.
61. See Fuller, *Civil-Military Conflict*, 196–206.
62. In the years after the Russo-Japanese War, the shortage of officers "reached critical proportions." Kenez, "A Profile of the Prerevolutionary Officer Corps," 129.
63. Michael Howard, "Men Against Fire: The Doctrine of the Offensive in 1914," 521.
64. Wildman, *The End of the Russian Imperial Army*, 39.
65. Steinberg, *All the Tsar's Men*, 5.
66. Rich, *The Tsar's Colonels*, 7.
67. Duz', *Istoriia vozdukhoplavaniia*, 326-38, 411–24.
68. Ibid., 408–24.
69. Such popular histories include, for example, V. I. Lavrenets, *Letchiki Rossii* (Moscow: Mashinostroenie, 1992) and G. V. Zalutskii, *Vydaiushchiesia russkie letchiki* (Moscow: Voennoe Izdatel'stvo, 1953).
70. For general treatments of the development of the tsarist air services, see David R. Jones, "The Beginnings of Russian Air Power"; Kilmarx, *A History of Soviet Air Power*; James Sterret, *Soviet Air Force Theory, 1918–1945* (New York: Routledge, 2007); Von Hardesty, "Early Flight in Russia"; Christopher C. Lovett, "Russian and Soviet Naval Aviation, 1908–1916" in *Russian Aviation and Air Power*: 108–25; Scott W. Palmer, *Dictatorship of the Air*; and Gregory Vitarbo, "'The Power, Strength, and Future of Russia': Aviation Culture and the Russian Imperial Officer Corps, 1908–1914" (Ph.D. Diss., University of Michigan, 1999). Soviet and Russian sources include Duz', *Istoriia vozdukhoplavaniia,* and L. V. Ruzov and IU. N. Iablochkin, *Gatchina. Istoricheskii ocherk*; and Lidiia Kuz'mina, *Pervye kryl'ia Rossii.*
71. This biographical information on Mikhail Efimov and Lieut. Nesterov is taken from various sources. Duz', *Istoriia vozdukhoplavaniia,* describes their activities at some length, while popular histories all include sections devoted to Efimov and Nesterov. See Lavrenets, *Letchiki rossii*; G. Zalutskii, *Vydaiushchiesia russkie letchiki*; and Tom Darcey, Alan Durkota, et al, *The Imperial Russian Air Service Famous Pilots and Aircraft of World War One* (Mountain View, CA: Flying Machines Press, 1996). As paragons of Russian aviation, they are also invariably discussed in memoirs of the period. Biographies specifically devoted to these men include E. V. Koroleva and V. A. Rudnik,

Soperniki orlov (Odessa: Maiak, 1974) for Efimov, while several choose Nesterov as their focus: V. Tkachev, *Russkii sokol* (Krasnodar, 1961); K. I. Trunov, *Petr Nesterov* (Moscow: Sovetskaia Rossiia, 1975); and K. E. Veigelin, *Put' letchika Nesterova* (Moscow: Gos. Izdatel'stvo Oborronnoi Promyshl., 1939).

Chapter 2

1. The Grand Duke's own relatively brief recollections of his extensive aviation activity are found in his memoirs: Aleksander Mikhailovich, *Once a Grand Duke* (New York: Cassell and Co. Limited, 1932), 237–23.
2. Robin Higham, *Air Power: A Concise History* (New Yok: St. Martin's, 1972).
3. For an excellent treatment of the relationship of aviation and the popular consciousness, see Robert Wohl, *A Passion for Wings: Aviation and the Western Imagination, 1908–1918* (New Haven: Yale University Press, 1994); also Stephen Kern, *The Culture of Time and Space 1880–1918* (Cambridge, MA: Harvard University Press, 1983).
4. This brief overview of the early years of Russian military aviation is taken from various sources: Duz', *Istoriia vozdukhoplavaniia*; Jones, "The Beginnings of Russian Air Power, 1907–22"; Kilmarx, *A History of Soviet Air Power*; Ruzov and Iablochkin, *Gatchina. Istoricheskii ocherk*; Stokesbury, *A Short History of Air Power*; and Morrow, *The Great War in the Air*. There is also some brief information on aviation in Menning, *Bayonets Before Bullets*.
5. Jones, "The Beginnings of Russian Air Power," 17.
6. Duz', *Istoriia vozdukhoplavaniia*, 227, 398–99.
7. Ruzov and Iablochkin, *Gatchina. Istoricheskii ocherk*, 60–64.
8. Russian State Military History Archive [RGVIA], f. 802, op. 3, d. 1023, ll. 67–69.
9. Ruzov and Iablochkin, *Gatchina. Istoricheskii ocherk*, 69.
10. As quoted in Jones, "The Beginnings of Russian Air Power," 17.
11. Ruzov and Iablochkin, *Gatchina. Istoricheskii ocherk*, 58; Duz', *Istoriia vozdukhoplavaniia,* 329. As the airplane was still so novel, it was thought better to procure and test different systems than to limit purchasing to one type at this early stage.
12. Jones, "The Beginnings of Russian Air Power," 17.
13. Duz', *Istoriia vozdukhoplavaniia*, 398–99.
14. Snyder, *The Ideology of the Offensive*, 41–56.
15. John H. Morrow, Jr., "Knights of the Sky: The Rise of Military Aviation," in *Authority, Identity, and the Social History of the Great War*, ed. Frans Coetzee and Marilyn Shevin-Coetzee (Providence: Berghahn Books, 1995), 307. See also Morrow, *The Great War in the Air*. For works which specifically address the French experience, see Edmond Petit, *La Vie quotidienne dans l'aviation en France au debut du XXe siecle (1900–1935)* (Paris: Hachette, 1977), and Charles Christienne, *A History of French Military Aviation* (Washington, D.C.: The Smithsonian Institution Press, 1986).
16. See Peter Fritzsche, *A Nation of Flyers: German Aviation and the Popular Imagination* (Cambridge, MA: Harvard University Press, 1992), chapter 1.
17. Fritzsche, *A Nation of Flyers*, 43.
18. Morrow, "Knights of the Sky," 309. For more on German military aviation, see Morrow, *Building German Airpower, 1909–1914* (Knoxville: University of Tennessee Press, 1976) and *German Airpower in World War I* (Lincoln: University of Nebraska Press, 1982).

19. See Avner Offer, *The First World War: An Agrarian Interpretation* (New York: Oxford University Press, 1989), and Offer, "The Working Classes, British Naval Plans and the Coming of the Great War," *Past and Present* 107 (May 1985): 204–26.
20. David Edgerton, *England and the Aeroplane: An Essay on a Militant and Technological Nation* (Basingstoke: Palgrave MacMillan, 1991). See also A. M. Gollin, *The Impact of Airpower on the British People and Their Government, 1909–1914* (Stanford: Stanford University Press, 1989) and Michael Paris, *Winged Warfare: The Literature and Theory of Aerial Warfare in Britain, 1859–1917* (New York: Manchester University Press, 1992).
21. See again Menning, *Bayonets Before Bullets*, chapters 6–7.
22. As quoted in Jones, "The Beginnings of Russian Air Power, 1907-1922," 16–17.
23. RGVIA, f. 802, op. 3, d. 1074, ll. 65–67.
24. "Aeroplany v budyshchei voine," *Birzhevie vedemosti*, (Sept. 14, 1910), 3.
25. "Vtoraia mezhdunarodnaia vystavka vozdukhoplavaniia v Parizhe," *Voennyi sbornik* vol. (January 1911), 172.
26. V. Bronskii, "Sport," *Voennyi sbornik* vol. 3 (March, 1911), 169.
27. Von Hardesty, "Early Flight in Russia," 11–12.
28. Imperatorskii Vserossiiskii Aeroklub, *Vozdukhoplavanie* (St. Petersburg, 1912), 34; also found in RGVIA, f. 2000, op. 7, d. 27, ll. 65–66.
29. S. A. B., "Vozdushnye grenadery," *Razvedchik* no. 1155 (Dec. 18, 1912), 858. Surely enough, in the fall of 1913, the Russian War Ministry would hold a performance competition for airplanes which would also include testing of equipment and designs for dropping bombs from aircraft. RGVIA, f. 2000, op. 7, d. 78, ll. 21.
30. RGVIA, f. 2000, op. 7, d. 44, ll. 14–16.
31. RGVIA, f. 802, op. 3, d. 1074, ll. 14–15.
32. This cannot be overstated, however, for more so than other nations, the Russian army from the start envisaged a serious contribution from soldier-pilots. This is an issue that will be explored in depth in chapter 7.
33. For a rosy picture from a contemporary regarding the unique suitability of the Russian political, social, and cultural system for the development of aviation, see Boris Tageev, *Aerial Russia: The Romance of the Giant Aeroplane* (New York: John Lane Co., 1916).
34. Duz', *Istoriia vozdukhoplavaniia*, 186.
35. For more detailed descriptions of these and other projects, see Duz', *Istoriia vozdukhoplavaniia,* and *Aviatsiia i vozdukhoplavanie v Rossii v 1907–1914 g.g. (Sbornik dokumentov)*, vols. 3–7 (Moscow: Akademiia Nauk S.S.S.R., 1970–77), an invaluable resource for the study of Russian aviation in this period. See also *Vozdukhoplavanie i aviatsiia v Rossii do 1907. Sbornik dokumentov i materialov*, (Moscow: Gos. Izdatel'stvo Oborronnoi Promyshl., 1956).
36. Again, notable among such officers was Capt. German of the OVS, the editor/publisher of *Vozdukhoplavatel'*, a journal published under the masthead of the Officers' Aeronautics School.
37. These organizations included the Imperial All-Russian Aeroclub, the Moscow Society of Aeronautics, the Odessa Aeroclub, and the Imperial All-Russian Technical Society.
38. RGVIA, f. 2000, op. 7, d. 44.
39. RGVIA, f. 802, op. 3, d. 1074, l 93. The letter was quoted at length here in a draft report of the Main Engineering Administration to the War Council.
40. "Feleton," *Razvedchik* no. 1049 (Dec. 7, 1910): 751–752.

41. The event was covered by all the major newspapers; Palmer, 'On Wings of Courage."
42. See again Palmer, "On Wings of Courage."
43. K. Bronskii, "Vpered!," *Voennyi sbornik* vol. 8 (August 1911), 185.
44. RGVIA, f. 802, op. 3, d. 1074, ll. 14–15.
45. *Sevastopol'skii aviatsionnyi illiustrirovannyi zhurnal* no. 1 (Nov. 19, 1910), 3.
46. Kovan'ko's clearest expression of his views came in his pamphlet published in 1912, *Thoughts on the Planned Organization of Military Aeronautics in Russia*. This document will be examined at length in chapter 4.
47. The article in question is also summarized and quoted at length in Koroleva and Rudnik, *Soperniki orlov*.
48. *Sevastopol'skii aviatsionnyi illiustrirovannyi zhurnal* no. 3 (December 5, 1910), 2.
49. IVAK, *Vozvanie*, 41. For some intriguing observations on the relationship of geography, power, and national identity in Russian thought, see Mark Bassin, "Russia Between Europe and Asia: The Ideological Construction of Geographical Space," *Slavic Review* 50, no. 1 (Spring 1991): 1–17.
50. Duz', *Istoriia vozdukhoplavaniia*, 71–79.
51. The regulations regarding these courses are reproduced in *Aviatsiia i vozdukhoplavanie v Rossii*, vol. 4, document no. 269: 4–5.
52. *Sevastopol'skii aviatsionnyi illiustrirovannyi zhurnal* nos. 42–43 (n.d., 1911), 2.
53. Higham, *Air power: A Concise History*, 2.
54. As quoted in Koroleva and Rudnik, *Soperniki orlov*, 69–70.
55. Ibid., 80–81.
56. A. Poltaratskii, "Vozdukhoplavanie," *Razvedchik* no. 1073 (May 24, 1911), 326.
57. One such set of instructions was published in *Razvedchik* no. 1062 (March 8, 1911), 139.
58. For more on this unique British experience, see Alfred Gollin, *No Longer An Island. Britain and the Wright Brothers, 1902–1909* (Stanford: Stanford University Press, 1984).
59. Imperatorskii Vserossiiskii Aeroklub, *Vozdukhoplavanie*, pp. 35–36, as found in RGVIA, f. 2000, op. 7, d. 27, l. 67.
60. For more on attitudes toward non-Russians see Mark von Hagen, "The Limits of Reform: The Multiethnic Imperial Army Confronts Nationalism, 1874–1917," in *Reforming the Tsar's Army*, 34–55, and Gregory Vitarbo, Nationality Policy and the Russian Imperial Officer Corps, 1905–1914," *Slavic Review* 66, 4 (Winter 2007): 682–701.
61. RGVIA, f. 802, op. 3, d. 1086, ll. 24–26.
62. Kenez, "A Profile of the Prerevolutionary Officer Corps," 136. See Denikin, *Staraia armiia*, vol. 2, 118.
63. James Stokesbury makes an analogous comment about Russian inventors and airplane construction: the small-scale, shop nature of the domestic aviation industry allowed room to experiment and diversify that would have been absent had a more developed, more centralized aviation industry been present, but this also entailed problems of limited production capacity and capital. This resulted in technical innovations that could not ultimately be developed to their full extent. Stokesbury, *A Short History of Air Power*, 21.
64. Durkota et al., *The Imperial Russian Air Service*, 201.
65. Zalutskii, *Vydaiushchiesia russkie letchiki*, 35-6.
66. See Morrow, *The Great War in the Air*, chapter 1, and Stokesbury, *A Short History of Air Power*, chapters 2–3.

Chapter 3

1. A. Poltaratskii, "Vozdukhoplavanie," *Razvedchik* no. 1073 (May 24, 1911): 323–24.
2. See Jones, "The Beginnings of Russian Air Power," 15–21; Kilmarx, *A History of Soviet Air Power*, 5–12; and Duz', *Istoriia vozdukhplavanii,* 326–332.
3. Many of these industrialists' appeals are preserved in *Aviatsiia i Vozdukhoplavanie v Rossii*, vols. 3–7.
4. Jones, "The Beginnings of Russian Air Power," 18–19.
5. *Vozdushnyi flot. Sil Rossii* (Moscow, 1913), 9.
6. Ibid., 9.
7. *Aviatsiia i vozdukhoplavanie v Rossii*, vol. 3, document no. 231, 117-20; also found in Rossiiskii Gosudarstvennyi Voenno-Istoricheskii Arkhiv (RGVIA), f. 802, op. 3, d. 1346, ll. 59–62.
8. Jones, "The Beginnings of Russian Air Power," 18–19.
9. N. E. Popov, *Voina i let voinov* (St. Petersburg, 1912), 55.
10. Ibid., 63.
11. Ibid., 71, 65.
12. "Aeroplany v budyshchei voine," *Birzhevie vedemosti*, (Sept. 14, 1910), 3.
13. The article in question is also summarized and quoted at length in Koroleva and Rudnik, *Soperniki orlov*.
14. Ofitserskaia Vozdukhoplavatel'naia Shkola, *Soobrazheniia po povodu planomernoi organizatsii voennago vozdukhoplavaniia v Rossii* (Saint Petersburg, 1912).
15. RGVIA, f. 2000, op. 7, d. 7, l. 49.
16. A. Kovan'ko, "Vozdushnyi Flot," in *Vozdushnyi flot- sil Rossii*, 17–23.
17. Ibid., 17.
18. Ibid., 21.
19. Ibid., 21–22.
20. Ibid., 22–23.
21. See for example Michael Adams, *The Great Adventure: Male Desire and the Coming of World War I* (Bloomington, 1990); Allen Frantzen, *Bloody Good: Chivalry, Sacrifice, and the Great* War (Chicago, 2003); Steven Clement, *For King and Kaiser! The Making of the Prussian Army Officer, 1860–1914* (New York: Greenwood Press, 1992).
22. This is not to suggest, however, that even in Britain, much less France and Germany, the "triumph" of capitalist values was unquestioned or absolute. See for example Martin Weiner, *English Culture and the Decline of the Industrial Spirit, 1850–1980* (New York: Cambridge University Press, 1981), and David Blackbourn and Geoff Eley, *The Peculiarities of German History: Bourgeois Societ and Politics in Nineteenth Century Germany* (New York: Oxford University Press, 1984).
23. James L. West, "Merchant Moscow in Historical Context," in *Merchant Moscow: Images of Russia's Vanished Bourgeoisie*, ed. J. West and Iu. Petrov (Princeton: Princeton University Press, 1998), 6.
24. Timothy McDaniel, *Autocracy, Capitalism, and Revolution in Russia* (Berkeley and Los Angeles: University of California Press, 1988), 16–23.
25. Thomas Owen, *Russian Corporate Capitalism from Peter the Great to Perestroika* (New York: Oxford University Press, 1995), 8–10.
26. Ibid., 116.
27. Ibid., 126–35.

28. McDaniel, *Autocracy, Capitalism, and Revolution in Russia*, 23.
29. Owen, *Russian Corporate Capitalism*, 116–120. See also for example Liah Greenfield, *Nationalism: Five Roads to Modernity* (Cambridge, MA: Harvard University Press, 1992), 189–274.
30. See again Scott Palmer, *Dictatorship of the Air*, pp. 11–75.
31. Regarding the popular response to aviation, see again Robert Wohl, *A Passion for Wings*; Steven Kern, *The Culture of Time and Space*; Alfred Gollin, *No Longer An Island*; Peter Fritzsche, *A Nation of Flyers: German Aviation and the Popular* Imagination (Cambridge, MA: Havard University Press, 1992); Joseph Corn, *Winged Gospel: America's Romance with Aviation, 1900*–1950 (New York: Oxford University Press, 1983); Bayla Singer, *Like Sex With Gods: An Unorthodox History of Flying* (College Station: Texas A&M University Press, 2003); and Laurence Goldstein, *The Flying Machine and Modern Literature* (Bloomington: Indiana University Press, 1986).
32. Wohl, *A Passion for Wings*, 5–30; Robin Higham, *Air Power: A Concise History* (London, 1972), 2.
33. John H. Morrow, "Knights of the Sky: The Rise of Military Aviation," 305–11.
34. These were included in pamphlets such as *Vozdushnyi flot: Sil Rossii* (Moscow, 1913), as well as regular coverage of the Grand Duke's activities and public pronouncements in *Sevastopolskii aviatsionnyi illiustrirovannyi zhurnal*.
35. Jones, "The Beginnings of Russian Air Power," 19.
36. Palmer, *Dictatorship of the Air*, 58–63.
37 Kilmarx, *A History of Soviet Air Power*, 8–9.
38. These publications included the essay "Vozdushnyi Flot," in *Vozdusnyi flot: sil Rossii*; his programmatic policy statement *Soobrazheniia po povodu voennago vozdukhoplavanii v Rossii* (St. Petersburg, 1912); and interviews for newspaper articles such as "Aeroplany v budyshchei voine," *Birzhevie vedemosti* (Sept. 14, 1910): 3. Unofficially, Kovan'ko was somewhat prolific in producing memorandums, reports, and letters to his superiors advancing and defending his policies.
39. See in particular the voluminous correspondence preserved in the primary source compilation *Aviatsiia i vozdukhoplavaniie v Rossii v 1907–1914 g.g.*.
40. Jones, "The Beginnings of Russian Air Power," 19.
41. See Palmer, *Dictatorship of the* Air, 49–50, and Vitarbo, "'The Power, Strength, and Future of Russia,'" chapters 2, 4.
42. Palmer, *Dictatorship of the Air*, 70–75; Vitarbo, "Military Aviation, National Identity, and the Imperatives of Modernity," 290; Jones, "The Beginnings of Russian Air Power," 18–20; Vitarbo, "'The Power, Strength, and Future of Russia,'" 194–245.
43. Such petitions and appeals litter the documentary records of the Officers Aeronautics School, the Department of the Air Force, the Main Engineering Administration, and the Aviation Section of the General Staff. See again RGVIA, fond 352, 844, 802, and 2000, respectively. A good number of such documents are also contained in the collection *Aviatsiia i Vozdukhoplavanie v Rossii*, vols. 3–7. For example, the OVF itself, despite its appeals for public donations, had received a 100,000 ruble annual subsidy.
44. *Aviatsiia I vozdukhoplavanie v Rossii*, vol. 4, no. 349, 147–49.
45. *Aviatsiia i vozdukhoplavanie v Rossii*, vol. 5, no. 452, 130.
46 Jones, "The Beginnings of Russian Air Power," 18–19.
47. RGVIA, f. 2000, op. 7, d. 177, ll. 110–11.
48. *Aviatsiia i vozdukhoplavanie v Rossii*, vol. 6, no. 528, 85.

49. Duz', *Istoriia vozdukhoplavaniia*, 286–358; Ruzov, *Gatchina. Istoricheskii ocherk*, 62.
50. This interview with Kovan'ko is cited in E.V. Koroleva and V.A. Rudnik, *Soperniki orlov,* 69–70.
51. M. Efimov, "Koe-chto ob aviatsii," *Sevastopol'skii aviatsionnyi illiustrirovannyi zhurnal* no. 41 (n.d., 1911), 5–6.
52. Only one pilot, Vasil'ev, actually reached Moscow. See the short account provided in "Perelet Peterburg-Moskva," *Razvedchik* no. 1081 (July 19, 1911), 447–49. The official account of the contest was published as *Otchet o pervom v Rossii perelete S.-Peterburg-Moskva. 10–15 iiulia 1911 g.* (St. Petersburg, 1911). For coverage of the withering criticism of the event on the part of the contemporary press, see also Palmer, *Dictatorship of the Air*, 37–41.
53. RGVIA, fond 2000, op. 7, delo 78, l. 1.
54. Ibid., l. 12.
55. Ibid., ll. 13, 14, 17.
56. The report, dated May 16, 1913, is found in RGVIA, f. 2000, op. 7, d. 78, ll. 91–2.
57. Kovan'ko may have been unduly pessimistic regarding the conditions of the flight, as the whole point of the contest was to keep repairs and outside aid and assistance to the pilots to a minimum; but his attitude was instructive. Nevertheless, 26 officers from other units did volunteer, and two were subsequently selected.
58. RGVIA, f. 802, op. 3, d. 1086, ll. 93.
59. Palmer, *Dictatorship of the Air*, 36; Vitarbo, "'The Power, Strength, and Future of Russia,'" 221–25, 254–60.
60. Palmer, *Dictatorship of the Air*, 48.
61. RGVIA, f. 2000, op. 7, d. 27, l. 4.
62. Kh.F. Stamat'ev, "Chem neobkhodimo rukovodstvovat'sia pri komplektovanii aviatsionnoi shkoli," *Sevastopolskii Aviatsionnyi Illustrirovannyi Zhurnal*, no. 8, January 9, 1911: 7–8.
63. N. Rozanov, "Shoffery aviatsii," *Sevastopolskii Aviatsionnyi Illustrirovannyi Zhurnal*, no. 25, May 29, 1911.
64. See again Owen, chs.1 and 5, And McDaniel, chs. 1–2.
65. See Duz', *Istoriia vozdukhoplavaniia*, chapter 7.
66. See Wohl, *A Passion for Wings,* chs. 2–3.

Chapter 4

1. N. Dukhanin, "O nashem voennom vozdukhoplavanii," *Razvedchik* 1057 (Feb. 1, 1911): 65–69.
2. This appraisal moved the editors to declare immediately in a footnote that "we completely disagree with the author in the assessment of the activity of this respected and highly useful institution."
3. N. Dukhanin, "K stat'e "O nashem voennom vozdukhoplavanii," *Razvedchik* 1060 (Feb. 22, 1911): 114.
4. Ibid., 114. It should be noted that this was not a problem for curmudgeons like Dukhanin alone. In a somewhat amusing sense, the activity of the OVF and particularly the role of Grand Duke Alexander Mikhailovich caused some problems of conscience as well for Soviet historians, who were usually in no rush to praise the efforts of the Imperial family. While aggressively questioning his motives, his competence, and his integrity, they

grudgingly admit and appreciate that he was at least doing *something* to build *otechestvennaia aviatsiia*, compared to what is described as the absolute inertia and apathy of the War Ministry. Duz', *Istoriia vozdukhoplavaniia*, 361–66; Koroleva and Rudnik, *Soperniki orlov*, 79–83.

5. Jones, "The Beginnings of Russian Air Power," 17–18.
6. RGVIA, f. 2000, op. 7, introduction.
7. *Aviatsiia i vozdukhoplavanie v Rossii*, vol. 3, document no. 205: 82–85.
8. *Aviatsiia i vozdukhoplavanie v Rossii*, vol. 3, document no. 245: 139–42.
9. Ibid., l. 141.
10. RGVIA, f. 2000, op. 7, d. 25, ll. 8–10.
11. Ibid., l. 8.
12. Ibid., l. 9.
13. Koroleva and Rudnik, *Soperniki orlov*, 79.
14. Ruzov and Iablokin, *Gatchina. Istoricheskii ocherk*, 62.
15. N. E. Popov, *Voina i let voinov*, (St. Petersburg, 1912), 64.
16. This telegram, dated Oct. 4, 1911, was reproduced in *Sevastopol'skii aviatsionnyi illiustrirovanyi zhurnal*, no. 39 (Nov. 3, 1911): 3.
17. Aleksandr Mikhailovich, *Once a Grand Duke*, 237.
18. As noted in Zhilinskii's report to Sukhomlinov of June 15, 1911. This report, which will be examined in more detail below, is found in *Aviatsiia i vozdukhoplavanie v Rossii*, vol. 3, document no. 205: 82–85.
19. A detailed account of this visit is found in *Sevastopol'skii aviatsionnyi illiustrirovanyi zhurnal* no. 39 (Nov. 3, 1911): 2–3.
20. RGVIA, f. 2000, op. 2, d. 242, l. 237.
21. Emphasis in the original by the author.
22. *Aviatsiia i vozdukhoplavanie v Rossii*, vol. 3, document no. 245, 140–41.
23. "Otchet o deiatel'nosti Vozdukhoplavatelnoi Chasti GUGSh za 1912 goda /s 11 Sentiabria po 1 Ianvaria 1913 goda/," found in RGVIA, f. 2000, op. 7, d. 44, ll. 13–14.
24. The letter, dated August 12, 1912, is found in RGVIA, f. 2000, op. 2, d. 1401, l. 12. The entire correspondence surrounding the inauspicious start of the Aeronautics Unit is found in ibid., ll. 1–14.
25. Ibid., l. 13.
26. The requisite War Department order and the corresponding regulations of the Aviation Section are found in *Aviatsiia i vozdukhoplavanie v Rossii*, vol. 4, document no. 335: 113–19.
27. The Aeronautics Committee, which handled such technical matters as establishing and verifying specifications for equipment and testing material that had been ordered, had also been transferred to the overall jurisdiction of the General Staff with the establishment of the Aeronautics Unit of GUGSh.
28. RGVIA, f. 844, op. 1, d. 37.
29. A regular theme in the reports of Gen. Shishkevich was frustration over the Duma's self-proclaimed recognition of the great national importance of military aviation and its tardiness in actually taking up requisite legislation and providing credits. This was especially true of his comprehensive report to Gen. Zhilinskii in Oct. 1912, a document which will be examined in greater depth below.
30. Duz', *Istoriia Vozdukhoplavaniia*, 361–62.
31. *Aviatsiia i vozdukhoplavanie v Rossii,* vol. 3, document no. 245: 139–42.

32. *Aviatsiia i vozdukhoplavanie v Rossii*, vol. 4, document no. 347: 135–46. The report can also be found in RGVIA, f. 2000, op. 7, d. 111, ll. 1–7.
33. Ibid., 144.
34. Ibid., 144–45.
35. *Aviatsiia i vozdukhoplavanie v Rossii,* vol. 5, document no. 403: 48–54. The report can also be found in RGVIA, f. 2000, op. 7, d. 231, ll. 11–14.
36. Ibid., 48.
37. Ibid., 51.
38. Ibid., 54.
39. A rough copy of this report is found in RGVIA, f. 2000, op. 7, d. 91, ll. 2–3.
40. Stokesbury, *A Short History of Air Power*, 18-19; Morrow, *The Great War in the Air*, 40–41.
41. Morrow, *The Great War in the Air*, 18–19.
42. See Matitiahu Mayzel, "The Formation of the Russian General Staff, 1880–1917: A Social Study," and "Generals and Revolutionaries, the Russian General Staff during the Revolution: A Study in the Transformation of a Military Elite." See also Steinberg, *The Tsar's Men*, and Van Dyke, *Russian Imperial Military Doctrine*.
43. Mayzel, "The Formation of the Russian General Staff," 307–08; see also Denikin, *Staraiia Armiia.*

Chapter 5

1. The experience of the "reforming officer" in the tsarist army is an issue that is difficult to precisely define or quantify. A reading of accounts from the period and subsequent memoirs leaves one with the conclusion that line officers who attempted in some way to modify or improve the rhythms and routines of army service were inevitably branded as "careerists" by a good many of their fellow officers. Such issues are also commented upon at length by Denikin, himself a reforming officer, in *Staraia armiia*. Though this is a subject that has not received a good deal of serious attention in the literature, for some comments see John Bushnell, "The Tsarist Officer Corps, 1881–1914: Customs, Duties, Inefficiency," and Allan Wildman, *The End of the Russian Imperial Army*, chapter 1.
2. For example, the editors of *Sevastopol Illustrated Aviation Journal* regularly applauded the energy and initiative of Col. Odintsov, as did the editors of the journal *Voennyi vozdushnyi flot,* published by the Grand Duke's Committee for Strengthening the Fleet. The renowned civilian aviator N. E. Popov also singled out Odintsov in his 1912 brochure *Voina i let voinov*, 71–73, while former soldier trainees at the school remembered Odintsov with admiration. See, for example, I. K. Spatarel', *Protiv chernogo barona* (Moscow: Voennoe Izdatel'stvo, 1967), a memoir from a soldier assigned to the Sevastopol School and one of the handful trained to fly there. Soviet biographies and popular histories likewise praise Col. Odintsov's efforts and ambitions. See Koroleva and Rudnik, *Soperniki orlov*; V. Puntus, *Kryl'ia kachi*; and V. Lavrenets, *Letchiki Rossii.*
3. From the letter of Vasilii Rebikov to Grand Duke Alexander Mikhailovich, RGVIA, f. 844, op. 1, d. 37. This conversation has been quoted at length in chapter 3.
4. RGVIA, f. 844, op. 1, d. 35.
5. Ibid., d. 35.
6. RGVIA, f. 802, op. 3, d. 1126, ll. 8–11.
7. N. Iatsuk, "Itogi goda," *Vozdukhoplavatel'* no. 1 (January 1912), 5.

8. From the letter of Vasilii Rebikov to the Grand Duke, RGVIA, f. 844, op. 1, d. 37.
9. RGVIA, f. 2000, op. 2, d. 242, l. 237. This letter is discussed at length in chapter 3.
10. Ibid., l. 237.
11. These tensions are apparent in the slew of reports, many of them concerning matters of expenditure and accounting, between the head of the OVF, Gen. Kaul'bars, the Grand Duke himself, and Col. Odintsov found in RGVIA, f. 844, op. 1, d. 36.
12. This affair comes under heavy criticism in Duz', *Istoriia vozdukhoplavaniia*, 365–66.
13. Ibid., 366.
14. RGVIA, f. 844, op. 1, d. 36, l. 5.
15. It was again the Aeronautics Section of the GIU which was liquidated and replaced by the Aeronautics Unit of GUGSh in September 1912.
16. RGVIA, f. 844, op. 1, d. 36, l. 7.
17. Boris Suvorin was the editor of both *Novoe vremia* and *Vechernee vremia*; Staff-Capt. Viktor- Berchenko was a flight instructor at the Sevastopol Aviation School; Capt. Prince Muruzi, as noted, was Col. Odintsov's assistant.
18. These articles apparently caught the interest of the Aeronautics Unit of GUGSh as well; a file in RGVIA, f. 2000, op. 7, d. 7, ll. 1–48, contains clippings of all of the newspaper articles cited below, as well as other articles concerning Russian aviation policies and personalities.
19. "Pis'mo ot Sevastopolia," *Rech'* (Sep. 8, 1912).
20. "Kto budet nachal'nikom Sevastopol'skoi shkoly aviatsii?" *Vechernee vremia* (Sep. 6, 1912).
21. *Novoe vremia* (Sep. 12, 1912).
22. RGVIA, f. 2000, op. 7, d. 82, l. 9
23. RGVIA, f. 2000, op. 7, d. 82, ll. 3–4.
24. Ibid., l. 4.
25. RGVIA, f. 2000, op. 7, d. 82, l. 20.
26. Rebikov's letter is again found in RGVIA, f. 844, op. 1, d. 37. The results of his meeting with Zhilinskii are discussed in chapter 3.
27. Ibid., d. 37
28. RGVIA, f. 2000, op. 7, d. 82, ll. 73–74.
29. Ibid., l. 74
30. Ibid., l. 75.
31. "Nasha voennaia aviatsiia," *Vechernee vremia* (June 15, 1913).
32. RGVIA, f. 2000, op. 7, d. 82, l. 56.
33. Ibid., l. 57.
34. Ibid., ll. 63–65.
35. Ibid., l. 64.
36. Ibid., l. 65.
37. See Fuller, *Civil-Military Conflict*, chapter 1.
38. Such issues are generally somewhat vexing ones for historians, for they are by definition murky and non-document based. Some factions and rivalries were well known, but it is often difficult for the contemporary historian to reconstruct at the level of the line all of the personal rivalries, minutiae of protocol, and points of honor that were seemingly so important to Russian officers. For some discussion of these issues see Fuller, *Civil-Military Conflict*, 3–46, 192–258, and Bushnell, "The Tsarist Officer Corps, 1881–1914." Fortunately, the officers of the aviation service were prolific both in their

rivalries and in their reports.

39. In some cases it was solved as much by luck and chance as any systematic policies.To note one example, the British air service was lucky to have British Major Hugh Trenchard. At age 39, he was only a few weeks short of the age limit for aviation service, yet he learned to fly in 13 days; for admittance to the army's Central Flying School one had to be a certified pilot already. Upon entry to the school he was immediately commissioned the school's permanent adjutant due to his combination of rank, experience, and pilot's diploma. Trenchard would go on to become the "father of the Royal Air Force" in World War I. Stokesbury, *A Short History of Air Power*, 23.
40. See for example his cover letter to War Minister Sukhomlinov regarding his pamphlet *Thoughts Regarding the Planned Organization of Military Aeronautics in Russia*, discussed at length in chapter 4.
41. For more on the role of the press and the public in the late Imperial period, see Louise McReynolds, *The News Under Russia's Old Regime: The Development of a Mass Circulation Press* (Princeton: Princeton University Press, 1991) and Jeffrey Brooks, *When Russia Learned to Read: Literacy and Popular Culture, 1861–1917* (Princeton: Princeton University Press, 1985).
42. See Menning, *Bayonets Before Bullets*, 221–22, and Fuller, *Civil-Military Conflict*, 237–44.
43. Duz', *Istoriia vozdukhoplavaniia*, 365–66.
44. Bernard Pares, *The Fall of the Russian Monarchy* (London, 1939), 109.

Chapter 6

1. For a general overview of such efforts, see John Morrow, *The Great War in the Air*; Morrow, "Knights of the Sky: The Rise of Military Aviation"; James Stokesbury, *A Short History of Air Power*; and Robin Higham, *Air power: A Concise History*.
2. A subsequent copy of this circular report, dated Dec. 15, 1910 and addressed to the Chief of Staff of the Omsk military district, is found in RGVIA, f. 802, op. 3, d. 1077, ll. 98–99.
3. In addition to the general medical criteria of the October 1910 circular, a War Department order of Oct. 11, 1911, provided a detailed list of all medical conditions that would disqualify officers for aviation service. These included a weak heart and circulatory disorders, "all forms" of hysteria and neurasthenia, traumatic and other forms of neurosis, convulsions or muscular tremors, alcoholism, chronic rheumatism, nasal disorders, and weak eyesight and hearing. *Aviatsiia i vozdukhoplavanie v Rossii*, vol. 3, document no. 232, 121–22.
4. A short piece in *Razvedchik* in the spring of 1911 announced that "as is known, the French War Ministry is paying particular attention to military aeronautics" and had thus recently announced its regulations for volunteers for aviation service; it reprinted these in full. "Inostrannyia armii," *Razvedchik* no. 1074 (May 31, 1911), 347.
5. Ibid., 267.
6. Such problems were only exacerbated as the War Ministry kept revising upward the planned number of aviation squadrons and the pace of their formation.
7. Capt. L. Sazerac de Forge, "Zhertvy aviatsii," translated by N. S. Aleksandrov, *Voennyi sbornik* vol. 1 (January, 1911): 173–80.
8. The commander of the Engineering Administration had thus ordered that 30 officers,

instead of the regular complement of 8, be assigned to train at the Aeronautical Training Park in 1908. Ibid., l. 145.

9. The lists and letters are found in RGVIA, f. 802, op. 3, d. 1126, ll. 88–91, 93, 95, and 96.
10. For Colonel Odintsov's report on this matter, see chapter 3.
11. Again, this contrast was made explicit in the satirical *Razvedchik* piece which juxtaposed the bureaucratic drudgery of a regimental chancellery and the opportunity for freedom offered by the airplane. "Feleton," *Razvedchik* no. 1049 (Dec. 7, 1910): 751–52.
12. RGVIA, f. 802, op. 3, d. 1077, l. 92.
13. Ibid., l. 111.
14. RGVIA, f. 802, op. 3, d. 1126, l. 3.
15. N. Dukhanin, "O nashem voennom vozdukhoplavanii," *Razvedchik* no. 1057 (Feb. 1, 1911), 68.
16. "Pravil'no-li?" *Razvedchik* no. 1176 (May 14, 1913): 307–08.
17. Ibid., 308.
18. These assertions surface to varying degrees throughout numerous works: V. Puntus, *Kryl'ia kachi* (Volgograd: Nizhne-Volzhskoe Kn. Izdatelstvo, 1970), a history of the Kachinskii Military Aviation School through both its Imperial (the Sevastopol Aviation School) and Soviet incarnations; Spatarel', *Protiv chernogo barona*; Koroleva and Rudnik, *Soperniki Orlov*; Lavrenets, *Letchiki Rossii*; Zalutskii, *Vydaiushchiesia russkie letchiki*; and Tkachev, *Russkii sokol*.
19. RGVIA, f. 802, op. 3, d. 1077, l. 69.
20. Ibid., l. 69.
21. RGVIA, f. 2000, op. 7, d. 95, l. 198.
22. RGVIA, f. 2000, op. 7, d. 95, l. 3.
23. RGVIA, f. 802, op. 3, d. 1077, l. 50.
24. Ibid., l. 50.
25. Ibid., l. 49.
26. The Aeronautics Unit was forced to respond to these complaints. In a letter of General Shishkevich to the Main Staff on Nov 19, 1912, he asked if it would be possible to modify the process of listing officers occupying vacancies in units. The requested changes, however, were apparently not made before the war. RGVIA, f. 2000, op. 7, d. 25, l. 57.
27. As these officers had been trained before the standardization of the requirements for the designation *voennyi letchik*, they received this designation retroactively by special dispensation. RGVIA, f. 2000, op. 7, d. 6, l. 47.
28. These lists are found in RGVIA, f. 2000, op. 7, d. 6, ll. 18, 58.
29. *Sevastopol'skii aviatsionnyi illiustrirovanyi zhurnal* no. 19 (April 10, 1911), 6.
30. The latter group simply had yet to take the exam. The lists, along with short attestations for each officer, are found in RGVIA, f. 2000, op. 7, d. 95, ll. 52–53.
31. RGVIA, f. 352, op. 2, d. 16, 39; f. 352, op. 3, d. 103; f. 802, op. 3, d. 1077, 1126, 1127; f. 2000, op. 7, d. 6, 27, 38, 72, 77, 91, 95; f. 844, op. 1, d. 26.
32. Recall from chapter 4 Lieut. Nesterov's unsuccessful attempts to petition the Main Engineering Administration for support and subsidies for his aircraft designs.
33. Lavrenets, *Letchiki Rossii*, 102.
34. Zalutskii, *Vydaiushchiesia russkie letchiki*, 37–38.
35. Zhilinskii's report, dated Oct. 15, 1912, is found in *Aviatsiia i vozdukhoplavanie v Rossii*, vol. 4, document no. 345, 132.
36. For example, in a document dated July 16, 1913, the Aeronautics Unit issued detailed

instructions for conducting exercises and flight operations at the aerodromes. *Aviatsiia i vozdukhoplavanie v Rossii*, vol. 5, document no. 423, 77–80. Such instructions had been worked out largely independently by the staffs of the two aviation schools, led by Col. Odintsov and Col. Ulianin. In 1912, the examination for the designation of *voennyi letchik* required a pilot to execute a flight of not less than 1 hour and 30 minutes, of which at least 30 minutes had to be at a height of more than 100 meters; to complete a gliding landing from a height of 100 meters; to perform assigned tasks of reconnaissance; and to pass a separate examination regarding the engine and technical aspects of the aircraft. Duz', *Istoriia vozdukhoplavaniia*, 400.

37. Such issues were indeed hotly debated at the time in both the specialty aeronautical press and popular newspapers throughout Europe and Russia. For some general comments about efforts to unravel the mysteries of the science and art of flight, see Higham, *Air Power: A Concise History*; Stokesbury, *A Short History of Air Power*; and Wohl, *A Passion for Wings*.
38. For indictments of such *beloruchki* in the Russian air services, see again Puntus, *Kryl'ia kachi*; Spatarel', *Protiv chernogo barona*; Koroleva and Rudnik, *Soperniki orlov*; Lavrenets, *Letchiki Rossii*; Zalutskii, *Vydaiushchiesia russkie letchiki*; and Tkachev, *Russkii sokol*.
39. M. Efimov, "Koe-chto ob aviatsii," *Sevastopol'skii aviatsionnyi illiustrirovannyi zhurnal* no. 40-41 (n.d., 1911), 5–6.
40. The general lack of staff turnover was cited by none other than Gen. Kovan'ko himself in his rebuttal to charges of nepotism. In regard to the Aviation Section of the OVS, Colonel Ulianin was its first and only commander in the pre-war period; Capt. Rudnev and Lieut. Kovan'ko, for example, both served 2–3 years as instructors, while the "*nesmeniamyi*" Gen. Kovan'ko was the only commander the OVS ever knew. In regard to the Sevastopol Aviation School, Capt., later Lieut. Col., Prince Muruzi, Capt. Viktor-Berchenko, and Mikhail Efimov all spent 2–3 years on the training staff.
41. For example, see the memoir of M. C. Machavariani, a Second Lieutenant assigned for practical training at the Aviation Section in May 1914. M. C. Machavariani, *Glaza na iug* (Tblisi: Merani, 1969), 1–9.
42. See for example Lavrenets, *Letchiki Rossii*, 140.
43. Duz', *Istoriia vozdukhoplavaniia*, 338–39.
44. For more detailed examination of this report see chapter 4.
45. Again, for more specific information regarding the work and activity of these officers, see Duz', *Istoriia vozdukhoplavaniia*, 299–301, 398–407. Documents regarding their plans and proposals can be found throughout the volumes of *Aviatsiia i vozdukhoplavanie v Rossii*. See for example the report of Chief of the General Staff Zhilinskii to the War Council, dated Dec. 22, 1913, recommending that they provided Nesterov with the necessary subsidies to perfect his experimental aircraft. Ibid., vol. 5, document no. 479, 166–67.
46. *Aviatsiia i vozdukhoplavanie v Rossii*, vol. 3, document no. 155, 4–5. The rules and regulations of the program of instruction are found in *Aviatsiia i vozdukhoplavanie v Rossii*, vol. 4, document no. 269, 4–5.
47. *Voennyi vozdushnyi flot* no. 3 (n.d.), pp. 15–16. For more on the activity of the OVF, see also "Otchet o deiatel'nosti komitet po usileniiu voennogo flota morskogo i vozdushnogo za 1913 g.," RGVIA, f. 2000, op. 7, d. 241.
48. Ibid., 16.

49. "Godovshchina ofitserskoi shkoly aviatsii O.V.F.," *Sevastopol'skii aviatsionnyi illiustrirovannyi zhurnal*, nos. 42–43 (n.d., 1911), 3.
50. For more detailed discussion of this incident, see chapter 5.
51. RGVIA, f. 2000, op. 7, d. 77, l. 4.
52. Ibid., l 5. His attestation reveals that Second Lieut. Iakov Treksin was from the 118th Shuiskii infantry regiment and had received his general education in the Zhitomir gymnasium and his military education in the Kazan military *uchilishche*, where he finished in the second rank.
53. This was a problem for all of the European air forces, owing to the relative dearth of accurate practical knowledge of aerodynamics and its effects on flight. See Stokesbury, *A Short History of Air Power*, ch. 1.
54. Koroleva and Rudnik, *Soperniki orlov*, 85.
55. These figures are drawn from the list of all Russian aviators, military and civilian, who perished in the period Oct. 1910–August 1914, provided in V. Lavrenets, *Letchiki Rossii*, appendix 3, 172–173.
56. N. Iatsuk, "Itogi goda," *Vozdukhoplavatel'* no. 1 (January 1912), 5. As the magazine was again published under the editorial masthead of the Officers' Aeronautics School, it should be noted that this was perhaps not an unbiased forum for such opinions.
57. Kilmarx, *A History of Soviet Air Power*, 11.
58. An excerpt of the actual report of this work, which was quite favorable, is found in *Aviatsiia i vozdukhoplavanie v Rossii*, vol. 5, document no. 428, 84–87.
59. The actual recruitment and training of observers in the early years of Russian military aviation, however, was somewhat haphazard; pilots, volunteer officers, and even soldiers were used as observers at the schools and in maneuvers on a largely trial and error basis.
60. RGVIA, f. 803, op. 3, d. 1123, ll. 14–15. A temporary detailed program of practicial and theoretical instruction and examinations for General Staff officers assigned to the aviation schools and units for training as observers was duly drawn up by Gen. Kovan'ko in spring 1911. A special examining commission of the OVS was to test these officers and review the training program upon their completion. RGVIA, f. 802, op. 3, d. 1123, l. 29, ll. 43–44.
61. Ibid., l. 15.
62. RGVIA, f. 802, op. 3, d. 1123, ll. 52–55.
63. Ibid., l. 52.
64. Ibid., l. 53. Col. Bulgarin noted that in the future a General Staff officer who had taken the course in the previous year could be assigned to training classes to perform this role.
65. Ibid., ll. 54–55.
66. Ibid., l. 55.
67. RGVIA, f. 2000, op. 7, d. 24, ll. 11–17.
68. According to information from the Sevastopol Aviation School, both officers had, in fact, been trained to fly, but how well or how often they did so is not clear.
69. Ibid., l. 13.
70. Ibid., l. 16.
71. Ibid., l. 17.
72. Reconstructing the daily experience of the line squadrons is even more difficult than recreating the environment of the two aviation schools. Formal reports and curt operational orders must be supplemented by sources that are frequently problematic, notably memoirs and popular histories. Most notable in this latter regard is the

pseudo-historical biography of Petr Nesterov by V. Tkachev, *Russkii sokol*. While this cannot be taken as an accurate history, when corroborated with other contemporary sources and memoirs it can be treated with care as a retrospective distillation of the challenges facing line aviation officers. Moreover, the retrospective picture painted by this and other works is supported at least obliquely by our knowledge of the tendencies of organizational behavior and the broader failings of human nature.

73. Again, these dilettantes and poseurs are favorite targets for the numerous memoirs and popular histories devoted to the true heroes—the serious aviators—of Russian Imperial military aviation.
74. Vozdukhoplavatel'naia rota, 7-aia, *Aeroplan i ego boevoe znachenie (k dokladu poruchika Mikos v ofitserskom garrizonnom sobranii 27 fev. 1912 g.)* (Kiev, 1912).
75. *Aviatsiia i vozdukhoplavanie v Rossii*, vol. 6, document no. 550, 129.
76. RGVIA, f. 2000, op. 2, d. 242, l. 237.
77. N. E. Popov, *Voina i let voinov*, 64.
78. "Sevastopol," *Sevastopol'skii aviatsionnyi illiustrirovannyi zhurnal* no. 12 (February 13, 1911), 2.
79. *Aviatsiia i vozdukhoplavanie v Rossii* vol. 3, document no. 245, 139–42.
80. RGVIA, f. 2000, op. 7, d. 82, l. 9.
81. This is true of both memoirs, Soviet and emigre, and Soviet historiography itself, which generally has nothing but fine things to say about Gen. Kovan'ko and his efforts to advance Russian aviation. See for example Duz', *Istoriia Vozdukhoplavaniia*, 422–24.
82. For example, see again the reports of Colonel Nemchenko regarding his mission to Western Europe in 1911-12. RGVIA, f. 802, op. 3, d. 1125, ll. 1-3, 5, 33, 73–75, 84–86, 87–88, 89–90, 91–99, 14, 24.
83. For example, the War Ministry held an aircraft contest in the fall of 1913, at which bomb-dropping devices were also tested and evaluated. RGVIA, f. 2000. op. 7, d. 78.
84. For a particularly controversial example, see the correspondence regarding the testing of a parachute designed by the Russian inventor Kotel'nikov. *Aviatsiia i vozdukhoplavanie v Rossii*, vol. 4, document nos. 348, 356, 363.
85. "Armeiskie zametki," *Razvedchik* no. 1230 (May 27, 1914): 334–35.
86. This play on words was an allusion to aviation terminology; the "death loop" (*mertvaia petlia*) was a loop performed in the air.
87. M. Grulev, "Den' vozdushnogo flota," *Razvedchik* no. 1231 (June 3, 1914): 349–50. The author noted that the Russian army was in a unique position to resolve this problem, unlike the French and German armies.
88. Ibid., 350.
89. *Aviatsiia i vozdukhoplavanie v Rossii*, vol. 4, document no. 347, 144–46.
90. See, for example, the 1913 program of demands and requirements by the War Ministry for construction and exploitation of aircraft. *Aviatsiia i vozdukhoplavanie v Rossii*, vol. 5, document no. 384: 23–24.
91. For good comparative general discussions of this issues, see Stokesbury, *A Short History of Air Power*, and Morrow, *The Great War in the Air*, ch. 1–2.
92. Morrow, *The Great War in the Air*, 84.
93. It is not a bit ironic that posterity subsequently ridiculed the disastrous selection of the Nieuports on primarily the same grounds cited in the article- that pilots of the line had not been sufficiently consulted while "dilettantes" had made the final choice.
94. Duz', *Istoriia vozdukhoplavaniia*, 366.

95. For example, the noted civilian aviator Princess Shakhovskaia declared in *Aero and Auto Life* that Nesterov's achievement was of little practical value for pilots.
96. "Mertvaia petlia na aeroplane," *Kievlianin* (Aug. 27, 1913), also contained in *Aviatsiia i vozdukhoplavanie v Rossii*, vol. 5, document. no. 438, 102.
97. Durkota et al., *The Imperial Russian Air Service*, 202.
98. Lieut. Col. Ts. Fedorov, "The First Aerial Ram," *Novoe russkoe slovo* (Nov. 23, 1944), 101. Unfortunately, no source is cited for this alleged interview. The article is contained amongst other papers with the unpublished typescript by Nikolai S. Karinsky, "Istoriia letnogo dela v Rossii do padeniia Imperii," (New York, 1944–45), 3 vols., contained in the Rare Books and Manuscripts Division of the New York Public Library.
99. Durkota et al., *The Imperial Russian Air Service*, 202.
100. This letter was reprinted in the Sept. 4 edition of *Kievlianin* and is contained in *Aviatsiia i vozdukhoplavanie v Rossii*, vol. 5, document no. 442: 108–09.
101. This is not to denigrate the talents and devotion of such officers. Numerous early graduates of the school and the staff itself went on to render productive service in the Russian air forces.
102. Class and competence, social pedigree and professional qualifications, need not be mutually exclusive. Echoing the arguments of military historians like Kenez and Wildman, Seymour Becker argues that regarding the professionalization of the Russian officer corps in the last decades of the Empire, nobles "may even be said to have turned it to their advantage by dominating the military schools that were so important a part of this growing professionalization." Becker, *Nobility and Privilege in Late Imperial Russia* (DeKalb, Ill., 1985), 110.
103. Zalutskii, *Vydaiushchiesia russkie letchiki*, 51.

Chapter 7

1. *Aviatsiia i vozdukhoplavanie v Rossii*, vol. 4, document no. 335: 113–18.
2. Ibid., 117. I. K. Spatarel', a mechanic later trained to fly as a soldier at the Sevastopol Aviation School, recalls that monthly flight pay for an officer pilot was 200 rubles, while that for a soldier pilot was 75. Spatarel', *Protiv chernogo barona*, 40.
3. Morrow, *The Great War in the Air*, 53.
4. RGVIA, f. 802, op. 3, d. 1126, ll. 29–30.
5. See Shishkevich's plan for the organization of aviation service examined in chapter 4.
6. RGVIA, f. 802, op. 3, d. 1126, l. 29. The report provided yearly projections in this regard: for the 19 aviation squadrons to be formed in 1912, 76 officer-pilots would be needed, 38 pilots from the ranks, and 95 specialists from the ranks; for the 22 squadrons to be formed in 1913 the numbers needed would be 88, 44, and 110 respectively; for the 12 squadrons to be formed in 1914, the necessary figures would be 48, 24, and 60.
7. Ibid., l. 29.
8. Ibid., l. 29. The report specified that the number of officers training at the OVS, which due to climate was able to graduate only one class per year, would be increased from the 15 provided for in the regulations to 35, while for the Sevastopol Aviation School, training two classes per year, the number would be increased from 40 to 60.
9. Ibid., l. 30.
10. RGVIA, f. 2000, op. 7, d. 6, l 1. The actual list is found on l. 18.
11. Ibid., l. 8.

12. *Aviatsiia i vozdukhoplavanie v Rossii*, vol. 5, document no. 403: 48–54.
13. Spatarel' notes that he and his handful of comrade soldier-pilots at the Sevastopol School all learned to fly in this way, aided by their own willingness to fly in unsuitable and even dangerous conditions that officers shunned. Spatarel', who no doubt did not have access to Engineering Administration reports during his tenure there, gives an alternate explanation for the expansion of training pilot soldiers: he notes that the handful who were first trained haphazardly flew so impeccably, despite the obstacles placed before them, that the command of the school decided to train more. Ibid., 34–35. Again, the fact that he seems to have not known that it was official policy to train soldier pilots bears witness to the ability of staff officers of the schools to subvert this policy. Supporting Spatarel's claims, Koroleva and Rudnik assert that it was indeed the Sevastopol School which was "presented with a fait accompli" of excellent soldier aviators, among them Spatarel', Semishkura, Kirsh, and Borovoi, and then responded by formally introducing the training of soldiers into the School's curriculum. *Soperniki orlov*, 110.
14. "Pervyi letchik iz riadovykh," *Sevastopol'skii Aviatsionnyi Illiustrirovannyi Zhurnal* no. 40–41 (n.d., 1911), 5.
15. Ibid., 5.
16. Morrow, *The Great War in the Air*, 53.
17. See again John Bushnell, "Peasants in Uniform: The Tsarist Army as a Peasant Society"; Bushnell, "The Tsarist Officer Corps, 1881–1914"; and Allan Wildman, *The End of the Russian Imperial Army*, chapter 1.
18. Spatarel', for example, argues that almost all of the aristocratic officers turned out to be *beloruchki,* while he asserts that in his tenure at the Sevastopol School he could only remember three officers who were seriously interested in mastering technical questions. *Protiv chernogo barona*, 21. Similar assertions are found in Koroleva and Rudnik, *Soperniki orlov*, 105–12.
19. Spatarel', *Protiv chernogo barona*, 22–30. Spatarel' also specifically cites Colonel Prince Muruzi, commander of the Sevastopol School, for treating mechanics well.
20. Spatarel' singles out for particular scorn among the instructors Lieut. Tunoshenskii, who had no desire to teach soldiers to fly, as well as Senior Lieut. Baron Buksgevden, who would regularly shout curses and epithets at his able and talented mechanic Emil Kirsh. He also cites Lieut. Il'in of the Life-Guards cavalry regiment, a son of a St. Petersburg factory owner, who distinguished himself by his arrogance. Ibid., 96.
21. Duz', *Istoriia vozdukhoplavaniia*, 328.
22. Spatarel', *Protiv chernogo barona*, 23–24. Soviet memoirs and histories in general display this same curious mix of awe, respect, and enmity for their adversary.
23. Ibid., 41. Of course, Spatarel' notes that such practices did not go over well with a great many of the aristocrat *beloruchki* aviators.
24. Durkota et al, *The Imperial Russian Air Service*, 237.
25. Again, see Bushnell, "The Tsarist Army as Peasant Society." Memoirs further attest that such master-serf attitudes were often most strong among the newly monied classes who affected the nobility's less charming tendencies, such as Spatarel's nemesis Lieut. Il'in.
26. RGVIA, f. 2000, op. 7, d. 95, l. 106. Lieut. Nesterov's attached service record indicated, among other things, that he was a hereditary noble from Kazan province and had been educated in the 2nd cadet corps and the Konstantinovskoe artillery *uchilishche*.
27. For a larger discussion of this project, see Gregory Vitarbo, "Nationality Policy and the Russian Imperial Officer Corps, 1905–1914."

28. RGVIA, f. 352, op. 2, d. 16, l. 13.
29. Ibid., l. 14.
30. Ibid., l. 16.
31. Ibid., l. 18.
32. Ibid., l. 20. A later draft letter, dated Nov. 28, 1913, from the Aeronautics Unit to the OVF noted that for other Catholic officers who had already been training in the schools when the restrictions had been issued, the Chief of the General Staff had decided to give them the choice of either serving in the Caucasus and Far East or returning to their regular unit. RGVIA, f. 2000, op. 7, d. 95, l. 74.
33. RGVIA, f. 2000, op. 7, d. 91, l. 137.
34. RGVIA, f. 2000, op. 7, d. 95, l. 69a. The letter of Gen. Artamanov to Gen. Shishkevich on behalf of Zagurskii is found on l. 71.
35. The excerpt was dated June 17, 1913, marked secret. RGVIA, f. 2000, op. 7, d. 82, l. 66.
36. "Korrespondentsiia *Razvedchika*," *Razvedchik* no. 1220 (March 18, 1914), 187–88.
37. Palmer, "On Wings of Courage."
38. "Sevastopol'skoe aviatsionnoe pole," *Razvedchik* no. 1098 (Nov. 15, 1911): 718–21.
39. *Aviatsiia i vozdukhoplavanie v Rossii*, vol. 5, document no. 457, 135–37.
40. Ibid., ocument no. 456,133–34.
41. Ibid., vol. 5, document no. 459, 139.
42. "Armeiskie zametki," *Razvedchik* no. 1208 (Dec. 24, 1913), 793.
43. See the War Ministry and General Staff reports and plans examined in chapters 3 and 4.
44. N. Grulev, "Den' Vozdushnogo Flota," *Razvedchik* no. 1231 (June 3, 1914): 349–50.
45. Ibid., 349.
46. RGVIA, f. 2000, op. 7, d. 95, l. 53.
47. RGVIA, f. 2000, op. 7, d. 77, ll. 1–2.
48. Koroleva and Rudnik, *Soperniki orlov*, 105.
49. Spatarel', *Protiv chernogo barona*, 30; Koroleva and Rudnik, *Soperniki orlov*, 109.
50. Koroleva and Rudnik, *Soperniki orlov*, 105–06.
51. V. Puntus, *Kryl'ia kachi*, 20.
52. See, for example, Gen. Zhilinskii's Dec. 1913 report to the War Council requesting that subsidies for Nesterov's experimental work be granted. The report noted that the goal of Nesterov's work—making takeoff and landing speeds slower and thus safer—was vitally important. *Aviatsiia i vozdukhoplavanie v Rossii*, vol. 5, document no. 479: 166–167.
53. Durkota et al, *The Imperial Russian Air Service*, 202.
54. The photograph is reproduced in both Veigelin, *Put' letchika Nesterova*, 87, and Zalutskii, *Vydaiushchiesia russkie letchiki*, 53. The many works which devote attention to Nesterov make a point of his respectful and progressive attitude toward his mechanics.
55. See, for example, the order of the Sevastopol School noting the July 1913 group flight of ten pilots from Sevastopol to Evpatoria, lead by Lieut. Tunoshenskii. *Aviatsiia i vozdukhoplavanie v Rossii*, vol. 5, document no. 419: 73.
56. M.C. Machavariani, *Glaza na iug*, 7–8.

Chapter 8

1. J. Glenn Gray, *The Warriors* (New York: Harcourt, Brace, 1959), 29.
2. Again, see Robert Wohl, *A Passion for Wings*.
3. *Sevastopol'skii aviatsionnyi illiustrirovannyi zhurnal* no. 1 (Nov. 19, 1910): 3–4.

4. Ibid., 4.
5. *Sevastopol'skii aviatsionnyi illiustrirovannyi zhurnal* no. 33 (August 17, 1911): 4–5.
6. Ibid., 4–5.
7. *Sevastopol'skii aviatsionnyi illiustrirovannyi zhurnal* no. 39 (Nov. 3, 1911): 1–2.
8. *Sevastopol'skii aviatsionnyi illiustrirovannyi zhurnal* nos. 42-43 (n.d., 1911): 2–6.
9. Ibid., 4–5.
10. Ibid., 5.
11. Ibid., 5.
12. RGVIA, f. 352, op. 2, d. 39, ll. 59–70, 73–74.
13. Ibid., l. 65.
14. These included Staff-Captains Datskevich, Dannilevskii, Boreiko, and Lieutenants Rudnev, Modrakh, Kovan'ko, Stoiakin, and Pankrat'ev, respectively. RGVIA, f. 352, op. 2, d. 39, l. 53.
15. RGVIA, f. 2000, op. 7, d. 91, l. 43.
16. More prosaically, German noted also that recognizing this date would also allow the staff, who had been in fact performing their duties from that day, to also draw pay from then.
17. Ibid., l. 40.
18. *Aero & avtomobil'naia zhizn'* no. 16 (August 15, 1913): 16–17.
19. Ibid., 16-17.
20. John Morrow notes that aviation funerary rituals performed the same function of linking military and public throughout the nations of Europe, prefiguring the wartime mourning of fallen heroes of the air. Morrow, *The Great War in the Air*, 34–35.
21. Palmer, "On Wings of Courage."
22. *Sevastopol'skii aviatsionnyi illiustrirovannyi zhurnal* no. 20 (April 22, 1911): 1–5.
23. According to the *Journal*, Staff-Capt. Matyevich-Matseevich, son of a lieutenant colonel, had been educated at the third Moscow cadet corps and the Nikolaevskii engineering *uchilishche*. In 1909 he had "brilliantly" finished the OVS, and in 1910 had been sent abroad to be trained on Bleriot aircraft. By agreement of the OVF and the Engineering Department, upon the opening of the Sevastopol Aviation School he was posted there as a supervising instructor. Ibid., 3–4.
24. Ibid., 5.
25. A. Poltaratskii, "Vozdukhoplavanie," *Razvedchik* no. 1073 (May 24, 1911): 323–26. He recalled that a near fatal mishap involving Matyevich at the First All-Russian Festival of Aeronautics had long gone unnoticed because it occurred almost simultaneously with the fatal crash of Capt. Matsievich.
26. A. Poltaratskii, "Vozdukhoplavanie," *Razvedchik* no. 1079 (July 5, 1911), 422.
27. *Sevastopol'skii aviatsionnyi illiustrirovannyi zhurnal* nos. 4–5 (52–53) (n.d., 1912): 1–6.
28. Ibid., 5.
29. Staff-Capt. Leon had finished the second Moscow cadet corps and the Mikhailovskoe artillery *uchilishche* in the first rank. He began service in the 16th mounted artillery battery in 1902; he arrived at the Sevastopol aviation school in July 1911 and had flown 3 hours and 4 minutes with his instructor, and 2 hours and 52 minutes independently. He passed the test for pilot on Nov. 28, 1911 and had flown as a pilot for an hour. Staff-Capt. Ianish, son of a retired major general, had finished the second cadet corps, then the Nikolaevskii engineering *uchilishche* in the first rank. After training in the officers' class of the Aeronautical Training Park, he served in the Siberian aeronautical battalion. He arrived at the Sevastopol aviation school on July 18, 1911. He had flown 1 hour and 33

minutes with his instructor and 2 hours 28 minutes independently, passing the test for pilot on Nov. 3, 1911. He had flown as a pilot for 43 minutes. Ibid., 4-5.

30. Ibid., 6.
31. *Sevastopol'skii aviatsionnyi illiustrirovannyi zhurnal* nos. 6–7, (n.d., 1912): 2–3.
32. Ibid., 2.
33. Ibid., 3.
34. "Pamiati voennago letchika Podporuchika Aleksandra Petrovicha Perlovskago," *Vozdukhoplavatel'* no. 5 (May 1913): 368–76.
35. Ibid., 371.
36. "Pamiati voennago letchika Poruchika M. G. Balabushka," *Vozdukhoplavatel'* no. 6 (June 1913): 460–64.
37. The son of an officer, Lieut. Balabushka had attended the Odessa cadet corps and the Nikolaevskoe engineering *uchilishche*, which he completed in 1909. Ibid.,460.
38. Ibid., 461.
39. Ibid., 463–64.
40. "Khronika," *Vozdukhoplavatel'* no. 10 (Oct. 1913), 728.
41. "Khronika," *Aero i avtomobil'naia zhizn'* no. 6 (March 1914): 20.
42. "Gibel' Poruch. Polikarpova i Mekhanika Drenova," *Vozdukhoplavatel'* no. 8 (August 1913): 604–06.
43. The obituary noted that Lieut. Polikarpov had attended the Konstantinovskoe artillery *uchilishche* , served in the 10th artillery brigade, and in the previous year had finished the Sevastopol Aviation School with the title of *voennyi letchik.* He had flown the craft in question three times; 24 years old, he had died a bachelor.
44. "Khronika," *Vozdukhoplavatel'* no. 8, (August 1913): 619–20.
45. *Russkii Invalid* correctly pointed out that by contractual obligations firms were obliged to train officers on aircraft intended to be introduced into the army's arsenal; thus, it noted, Russian officers regularly trained on such craft both at home and abroad.
46. Ibid., 620.
47. *Sevastopol'skii aviatsionnyi illiustrirovannyi zhurnal* nos. 4–5 (52–53) (n.d., 1912): 4.
48. N. Anoshchenko, "Slava pogibshim geroiam!" *Aero i Avtomobil'naia Zhizn'* no. 13 (July 1, 1914): 7–10.
49. The article provided a list of all dead Russian aviators, starting with Capt. Matsievich in 1910 and continuing to June 1914. The list included 32 army officers and soldiers; the amount of crashes, as pointed out, had risen alarmingly in 1913 and 1914. Ibid., 10.
50. See the accounts, for example, in Spatarel', *Protiv chernogo barona*, 29–30, and Koroleva & Rudnik, *Soperniki orlov*, 109.

Conclusion

1. Jones, "The Beginnings of Russian Air Power," 20, 22; Duz', *Istoriia vozdukhoplavaniia*, 202, 413.
2. A. P. Zhilin, "Bol'shaia programma po usileniiu russkoi armii," *Voenno-istoricheskii zhurnal* no. 7 (July 1974), 95.
3. Jones, "The Beginnings of Russian Air Power," 19.
4. A. Kovan'ko, "Vozdushnyi Flot," *Vozdushnyi flot—sil Rossii*, 23.
5. Duz', *Istoriia vozdukhoplavaniia*, 411-12; Morrow, *The Great War in the Air*, 47.
6. "Khronika," *Razvedchik* no. 1212 (Jan. 21, 1914): 45–46. A scathing critique of the

record of the General Staff in the sphere of aviation which appeared in *Golos moskvy* upon the occasion of the liquidation of the Aeronautics Unit was reprinted in *Razvedchik* no. 1214 (Feb. 4, 1914): 93; in reply, the editors could only lamely ask rhetorically if the War Department would indeed announce in the newspapers all of its efforts in this field.

7. "Sevastopol'," *Sevastopolskii Aviatsionnyi Illiustrirovannyi Zhurnal* no. 2 (Feb. 13, 1911), 2.
8. See again John Morrow, *The Great War in the Air*, chapter 1.
9. Harley Balzer, ed., *Russia's Missing Middle Class:The Professions in Russian History* (Armonk & New York: M.E. Sharpe, 1996), 21. He is one of many commentators to address professionalization as a function of modernity. See Edith Clowes, Samuel Kassow, and James West, eds. *Between Tsar and People: Educated Society and the Quest for Public Identity in Late Imperial Russia* (Princeton: Princeton University Press, 1991).
10. The growing tensions between military professionals and other constituencies in the state apparatus were largely over the conditions of their service and their feeling that the state was not being solicitous enough to their needs. See Fuller, *Civil-Military Conflict.*
11. Ibid., *Civil-Military Conflict*, 203.
12. A. Kovan'ko, "Vozdushnyi Flot," *Vozdushnyi flot- sil Rossii*, 22.
13. Pintner cites these factors as well as the army's decision to keep large numbers of soldiers in the standing army despite the existence of a large reserve. See Pintner, "The Burden of Defense in Imperial Russia, 1725-1914." For more analysis of the larger and long-term constraints facing the Russian army, see Fuller, *Strategy and Power in Russia.*
14. Jones, "The Beginnings of Russian Air Power," 21–23.
15. Zalutskii, *Vydaiushchiesia russkie letchiki*, 30.

| Bibliography

Archival Sources

Russian State Military History Archive [RGVIA]
fond 352. Officers' Aeronautics School
fond 2000. Main Administration of the General Staff
fond 802. Main Engineering Administration
fond 844. Department of the Air Force
fond 400. Main Staff

Hoover Institution Archives, Stanford University
M. Korol'kov collection
E. A. Nikol'skii collection

Bakhmeteff Archive, Columbia University
V. Dreier collection
I. Grebenshchikov collection
K. Khagondakov collection

Primary Sources

Aviatsiia i vozdukhoplavanie v Rossii v 1907–1914 g.g. (Sbornik dokumentov). vols. 3–7. Moscow: Akademiia Nauk S.S.S.R., 1970–1977.

Aviatsiia v Rossii: [Sbornik]. K 100-letiiu otech. samoletostroeniia. Moscow, 1983.

Imperatorskii Vserossiiskii Aeroklub. *Vozdukhoplavanie.* St. Petersburg, 1912.

Kryl'ia rodiny: /Sbornik/. Moscow, 1983.

Ofitserskaia Vozdukhoplavatel'naia Shkola, *Soobrazheniia po povodu planomernoi organizatsii voennago vozdukhoplavaniia v Rossii.* St. Petersburg: tip. M. V. Leon'teva, 1912.

Otchet Odesskago Aerokluba. s 11 marta 1908 g. po 10 maia 1910 g. Odessa, 1910.

Otchet o pervom v Rossii perelete S.-Peterburg-Moskva. 10–15 iiulia 1911 g.. St. Petersburg, 1911.

Popov, N. E. *Voina i let voinov.* St. Petersburg, 1912.

Rodnykh, Aleksandr. *Istoriia vozdukhoplavaniia i letaniia v Rossii.* vol. 1. St. Petersburg, 1912.

———. *Voina v vozdukhe v byloe vremia i teper'.* vol. 1. Petrograd, 1915.

Tageev, Boris L. *Aerial Russia: The Romance of the Giant Aeroplane.* London: John Lane Co., 1916.

Vozdukhoplavanie i aviatsiia v Rossii do 1907. Sbornik dokumentov i materialov. Moscow: Gos. Izdatel'svo Oboporonnoi Promyshl., 1956.

Vozdukhoplavatel'naia rota, 7-aia. *Aeroplan i ego boevoe znachenie. (k dokladu poruchika Mikos v ofitserskom garrizonnom sobranii 27 fev. 1912 g.).* Kiev, 1912.

Vozdushnyi flot- sil Rossii. Moscow, 1913.

Journals

Aero i avtomobil'naia zhizn'
Birzhevye vedomosti
Golos moskvy
Kievlianin
Novoe vremia
Razvedchik
Rech'
Russkii invalid
Sevastopol'skii aviatsionnyi illiustrirovannyi zhurnal
Vechernee vremia
Voennyi sbornik
Voennyi vozdushnyi Flot
Vozdukhoplavatel'

Secondary Sources

Aleksandr Mikhailovich. *Once a Grand Duke*. New York: Cassell and Co. Limited, 1932.
Aviatsiia Entsiklopediia. Moscow, 1994.
Balzer, Harley, ed. *Russia's Missing Middle Class: The Professions in Russian History*. Armonk & New York: M..E. Sharpe, 1996.
Bassin, Mark. "Russia Between Europe and Asia: The Ideological Construction of Geographical Space." *Slavic Review* 50 no. 1 (Spring 1991): 1–17.
Becker, Seymour. *Nobility and Privelege in Late Imperial Russia*. Dekalb, IL: Northern Illinois University Press, 1985.
Beskrovnyi, L.G. *Armiia i flot Rossii v nachale XX veka: Ocherki voenno-ekonomicheskogo potentsiala*. Moscow: Nauka, 1986.
———, ed. *Russkaia voennaia teoreticheskaia mysl' XIX i nachala XX vekov*. Moscow: Nauka, 1960.
Blackbourn, David and Geoff Eley, eds. *The Peculiarities of German History: Bourgeois Society and Politics in Nineteenth Century Germany*. New York: Oxford University Press, 1984.
Boyd, Alexander. *The Soviet Air Force Since 1918*. New York: Stein and Day, 1977.
Brooks, E. Willis. "Reform in the Russian Army, 1856-1861," *Slavic Review* 43 (Spring 1984): 63–82.
Brooks, Jeffrey. *When Russia Learned to Read: Literacy and Popular Culture, 1863–1917*. Princeton: Princeton University Press, 1985.
Bushnell, John. *Mutiny Amid Repression*. Bloomington: Indiana University Press, 1985.
———. "Peasants in Uniform: The Tsarist Army as a Peasant Society," *Journal of Social History* 13, no. 4 (Summer 1980): 565–576.
———. "The Tsarist Officer Corps, 1881–1914: Customs, Duties, Inefficiency," *American Historical Review* 86, no. 4 (October 1981): 753–780.
Challener, Richard. *The French Theory of the Nation in Arms, 1866–1939*. New York: Russell and Russell, 1965.
Christienne, Charles. *A History of French Military Aviation*. Washington, D.C.: Smithsonian Institution Press, 1986.
Clemente, Steven E. *For King and Kaiser! The Making of the Prussian Army Officer, 1860–1914*. New York: Greenwood Press, 1992.

Clowes, Edith, Samuel D. Kassow, and James L. West, eds. *Between Tsar and People: Educated Society and the Quest for Public Identity in Late Imperial Russia.* Princeton: Princeton University Press, 1991.

Cochrane, Dorothy, Von Hardesty, and Russell Lee. *The Aviation Careers of Igor Sikorsky.* Seattle: University of Washington Press, 1989.

Coetzee, Frans and Marilyn Shevin-Coetzee, eds. *Authority, Identity, and the Social History of the Great War.* Providence, RI: Berghahn Books, 1995.

Coffman, Edward. "The Long Shadow of The Soldier and the State," *Journal of Military History* 55 (January 1991): 69–82.

Collier, Basil. *A History of Air power.* 1st American ed. New York: MacMillan, 1974.

Corn, Joseph. J. *The Winged Gospel: America's Romance with Aviation, 1900–1950.* New York: Oxford University Press, 1983.

Craig, Gordon. *The Politics of the Prussian Army, 1640–1945.* New York: Oxford University Press, 1964.

Crummey, Robert, ed. *Reform in Russia and the U.S.S.R.: Past and Prospects.* Urbana, IL: University of Illinois Press, 1989.

Darcey, Tom, Durkota, Alan, et al, *The Imperial Russian Air Service Famous Pilots and Aircraft of World War One.* Mountain View, CA: Flying Machines Press, 1996.

Deak, Istvan. *Beyond Nationalism: A Social and Political History of the Hapsburg Officer Corps.* New York: Oxford University Press, 1990.

Demeter, Karl. *The German Officer-Corps in Society and State, 1650–1945.* New York: Weidenfeld and Nicolson, 1965.

Denikin, Anton I. *The Career of a Tsarist Officer: Memoirs 1872–1916.* Translated by Margaret Potoski. Minneapolis: University of Minnesota Press, 1975.

———. *Staraiia armiia.* 2 vols. Paris: Rodnik, 1929–1931.

Donne, Michael. *Per ardua ad astra: Seventy Years of the RFC and RAF.* New York: Harper Collins, 1984.

Duz', Petr D. *Istoriia vozdukhoplavaniia i aviatsii v Rossii: (Period do 1914 g.).*, 2nd ed. Moscow: Mashinostroenie, 1995.

———. *Istoriia vozdukhoplavaniia i aviatsii v Rossii: iiul' 1914 g.–oktiabr' 1917 g.*, 2nd ed. Moscow: Mashinostroenie, 1986.

Ecksteins, Modris. *Rites of Spring: The Great War and the Birth of the Modern Age.* New York: Doubleday, 1989.

Edgerton, David. *England and the Aeroplane: An Essay on a Militant and Technological Nation.* Basingstoke: Palgrave MacMillan, 1991.

Erickson, John. *The Soviet High Command: A Military-Political History, 1918–1941*, 3rd ed. New York: Routledge, 2001.

Finer, Samuel. *The Man on Horseback: The Role of the Military in Politics.* London: Pall Mall Press, 1962.

Finne, K. N. *Igor Sikorsky, The Russian Years.* Edited by Carl J. Bobrow and Von Hardesty. Translated by Von Hardesty. Washington, D.C.: The Smithsonian Institution Press, 1987.

Freeze, Gregory L. "The Soslovie (Estate) Paradigm and Russian Social History." *American Historical Review* 91 no. 1 (1986): 11–36.

Fritzsche, Peter. *A Nation of Fliers: German Aviation and the Popular Imagination.* Cambridge, MA: Harvard University Press, 1992.

Fuller, William C. *Civil-Military Conflict in Imperial Russia 1881–1914.* Princeton: Princeton University Press, 1985.

———. *Strategy and Power in Russia, 1600–1914.* New York: The Free Press, 1992.

Gatrell, Peter. "After Tsushima: Economic and Administrative Aspects of Russian Naval Rearmament, 1905–1913," *Economic History Review* 43, no. 2 (May 1990): 255–270.

———. *Government, Industry, and Rearmament in Russia, 1900–1914: The Last Argument of Tsarism*. New York: Cambridge University Press, 1994.

Gerschenkron, Alexander. *Economic Backwardness in Historical Perspective: A Book of Essays*. Cambridge, MA: Harvard University Press, 1962.

Gibbs-Smith, Charles H. *The Aeroplane: An Historical Survey of its Origins and Development*. London: H.M. Stationery, 1960.

Goldstein, Edward Ralph. "Military Aspects of Russian Industrialization: The Defense Industries, 1890–1917." Ph.D. diss., Case Western Reserve University, Cleveland, 1971.

Goldstein, Laurence. *The Flying Machine and Modern Literature*. Bloomington: Indiana University Press, 1986.

Gollin, A. M. *The Impact of Air Power on the British People and Their Government, 1909–1914*. Stanford: Stanford University Press, 1989.

———. *No Longer an Island. Britain and the Wright Brothers, 1902–1909*. Stanford: Stanford University Press, 1984.

Golovin, N. N. *The Russian Army in the World War*. New Haven, Conn., 1931.

Gray, J. Glenn. *The Warriors: Reflections on Men in Battle*. New York: Harcourt, Brace, 1959.

Greenfield, Liah. *Nationalism: Five Roads to Modernity*. Cambridge, MA: Harvard University Press, 1992.

Hardesty, Von. "Early Flight in Russia." In *Russian Aviation and Air Power in the Twentieth Century*, edited by Robin Higham et al., 18-36. Portland, OR: Frank Cass, 1998.

Higham, Robin D. *Air power: A Concise History*. New York: St. Martin's, 1972.

Higham, Robin and Jacob W. Kipp, eds. *Soviet Aviation and Airpower: A Historical View*. Boulder, CO: Westview, 1977.

Higham, Robin, John T. Greenwood, and Von Hardesty, eds., *Russian Aviation and Air Power in the Twentieth Century*. Portland, OR: Frank Cass, 1998.

Howard, Michael. "Men Against Fire: The Doctrine of the Offensive in 1914." In *Makers of Modern Strategy from Machiavelli to the Nuclear Age*, edited by Peter Paret, 510–526. Princeton: Princeton University Press, 1986.

Huntington, Samuel. *The Soldier and the State*. Cambridge, MA: Harvard University Press, 1957.

Hutchinson, J. F. "The Octobrists and the Future of Imperial Russia as a Great Power," *Slavonic and East European Review* 50, no. 119 (April 1972): 220–237.

Janowitz, Morris. *The Professional Soldier, A Social and Political Portrait*. Glencoe, IL: Free Press, 1960.

Jones, David. *The Military-Naval Encyclopedia of Russia and the Soviet Union*. Gulf Breeze, FL: Academic International Press, 1984.

Kagan, Frederick W. and Robin Higham, eds. *The Military History of Tsarist Russia*. New York: Palgrave, 2002.

Karinsky, Nikolai S. "Istoriia letnogo dela v Rossii do padenii Imperii" (New York, 1944–45), unpublished manuscript.

Kavtaradze, A. G. *Voennie spetsialisty na sluzhbe Respubliki Sovetov 1917–1920 g.g.*. Moscow: Nauka, 1988.

Keep, John L. H. *Soldiers of the Tsar: Army and Society in Russia, 1462–1874*. New York: Oxford University Press, 1985.

Keithly, David. "War Planning and the Otbreak of War in 1914," *Armed Forces and Society* 12, (Summer 1986): 553–579.

Kenez, Peter. "A Profile of the Prerevolutionary Officer Corps," *California Slavic Studies* 7, (1973): 121–158.

———. "Russian Soldiers and Military Conflicts," *Slavic Review* 45 (Winter 1986): 720–723.

Kern, Stephen. *The Culture of Time and Space 1880–1918*. Cambridge, MA: Harvard University Press, 1986.

Kersnovskii, A. A. *Istoriia russkoi armii*. Moscow: Golos, 1992.

Kilmarx, Robert A. *A History of Soviet Air Power*. New York: Praeger, 1962.

Kipp, Jacob W. "The Russian Navy and Technological Transfer." In *Russia's Great Reforms, 1855–1881*, edited by Ben Ekloff, John Bushnell, and Larissa Zakharova. Bloomington, IN: Indian University Press, 1994.

———. "Soldiers and Civilians Confronting Future War: Lev Tolstoy, Jan Bloch and Their Russian Military Critics." In *Tooling for War: Military Transformation in the Industrial Age*, edited by Stephen D. Chiabotti, 189–203. Chicago: Imprint Publications, 1996.

Koroleva, E. V. and V. A. Rudnik. *Soperniki orlov*. Odessa: Maiak, 1974.

Krasovskii, Stepan A. *Zhizn' v aviatsii*. Moscow: Voenizdat, 1968.

Kuprin, Aleksandr. *Sobranie sochinenii*. Moscow: Pravda, 1964.

Kuz'mina, Lidiia M. *Pervye kryl'ia Rossii: Stranitsy vekovoi istorii aviapredpriiatiia*. Moscow: Voennoe izdatel'stvo, 1993.

Lavrenets, Vsevolod I. *Letchiki Rossii*. Moscow: Mashinostroenie 1992.

Lebow, Eileen. *Before Amelia: Women Pilots in the Early Days of Aviation*. Washington. D.C.: Brassey's, Inc., 2002.

Lee, Asher, ed. *The Soviet Air and Rocket Forces*. Westport, CT: Greenwood Press Reprint, 1959.

Lieven, Dominic. *Empire: The Russian Empire and its Rivals*. New Haven: Yale University Press, 2002.

———. *Russia and the Origins of the First World War*. New York: St. Martin's Press, 1983.

Lincoln, W. Bruce. *In War's Dark Shadow: The Russians Before the Great War*. New York: Dial Press, 1983.

Luvaas, Jay. *The Military Legacy of the Civil War: The European Inheritance*. Chicago: University of Chicago Press, 1959.

Lyons, M. *The Russian Imperial Army: A Bibliography of Regimental Histories and Related Works*. Stanford: Hoover Institution Press, 1968.

Machavariani, M. C. *Glaza na iug*. Tblisi: Merani, 1969.

Mayzel, Matitiahu. "The Formation of the Russian General Staff, 1880-1917: A Social Study." *Cahiers du Monde russe et sovietique* 16, nos. 3–4 (July/Dec. 1975): 297–320.

———. "Generals and Revolutionaries, the Russian Genera Staff during the Revolution: A Study in the Transformation of a Military Elite." In *Studien zur Militargeschichte Militarwissenschaft und Konfliktforschung*, vol. 19. Osnabruck, 1979.

McDaniel, Tim. *Autocracy, Captitalism, and Revolution in Russia*. Berkeley and Los Angeles: University of California Press, 1988.

McReynolds, Louise. *The News Under Russia's Old Regime: The Development of a Mass Circulation Press*. Princeton: Princeton University Press, 1991.

Menning, Bruce. *Bayonets Before Bullets*. Bloomington: Indiana University Press, 1986.

Military History Symposium (U.S.) (12th: 1986: United States Air Force Academy). *Transformation in Russian and Soviet Military History: Proceedings of the Twelfth Military History Symposium, United States Air Force Academy, 1–3 October 1986*. ed. Col. Carl W. Reddel, USAF. Washington, D.C., 1990.

Miller, Forrest. *Dmitrii Miliutin and the Reform Era in Russia*. Vanderbilt: Vanderbilt

University Press, 1968.
Mondey, David, ed. *The International Encyclopedia of Aviation.* New York: Crown Publishers, 1986 .
Morrow, John H. *Building German Airpower, 1909–1914.* Knoxville: University of Tenessee Press, 1976.
———. *German Air Power in World War I.* Lincoln: University of Nebraska Press, 1982.
———. *The Great War in the Air: Military Aviation from 1909 to 1921.* Washington, D.C.: The Smithsonian Institution Press, 1993.
Mosse, George. *The Crisis of German Ideology: Inytellectual Origins of the Third Reich.* New York: Grosset and Dunlap, 1964.
Offer, Avner. *The First World War: An Agrarian Interpretation.* New York: Oxford University Press, 1989.
———. "The Working Classes, British Naval Plans and the Coming of the Great War." *Past and Present* 107 (May 1985): 204–226.
Owen, Thomas C. *Russian Corporate Capitalism from Peter the Great to Perestroika.* New York: Oxford University Press, 1995.
Palmer, Scott W. *Dictatorship of the Air: Aviation Culture and the Fate of Modern Russia.* New York: Cambridge University Press, 2006.
———. "On Wings of Courage: Public 'Air-Mindedness' and National Identity in Late Imperial Russia." *Russian Review* 54, 2 (April 1995): 209–226.
Paret, Paret, ed. *Makers of Modern Strategy from Machiavelli to the Nuclear Age.* Princeton: Princeton University Press, 1986.
Paris, Michael. *From the Wright Brothers to Top Gun: Aviation, Nationalism, and Popular Cinema.* New York: Manchester University Press, 1995.
———. *Winged Warfare: The Literature and Theory of Aerial Warfare in Brtain, 1859–1917.* New York: Manchester University Press, 1992.
Parramore, Thomas. *First to Fly: North Carolina and the Beginnings of Aviation.* Chapel Hill: University of North Carolina Press, 2002.
Penrose, Harald. *British Aviation: the Pioneer Years 1903–1914.* London: Aero Publishers, 1967.
Petit, Edmond. *La vie quotidienne dans l'aviation en France au debut du XXe siecle, 1900–1935.* Paris: Hachette, 1977.
Perrins, Michael. "Army Politics and Russian Society 1462–1937," *European History Quarterly* 17 (April 1987): 219–223.
Pintner, Walter. "The Burden of Defense in Imperial Russia, 1725–1914," *Russian Review* 43 (July 1984): 231–259.
———. "The Imperial Russian Army: Society, Politics, Economics, and Revolution," *Russian Review* 46 (October 1987): 439–446.
———. "Russian Military Thought: The Western Model and the Shadow of Suvorov." In *Makers of Modern Strategy from Machiavelli to the Nuclear Age*, edited by Peter Paret, 354–75. Princeton: Princeton University Press, 1986.
Pintner, Walter and Daniel K. Rowney, eds. *Russian Officialdom. The Bureaucratization of Russian Society from the Seventeenth to the Twentieth Century.* Chapel Hill: University of North Carolina Press, 1980.
Ponomareff, Dmitry. "Political Loyalty and Social Composition of the Military Elite: The Russian Officer Corps, 1861–1903." RAND Paper, ser. P-6052 (Nov. 1977).
Puntus, V. *Kryl'ia kachi.* Volgograd: Nizhne-Volzhskoe Kn. Izdatel'stvo, 1970.
Riaboff, Alexander. *Gatchina Days: Reminiscences of a Russian Pilot.* Edited by Von

Hardesty. Washington, D.C.: The Smithsonian Institution Press, 1986.
Rich, David. *The Tsar's Colonels: Professionalism, Strategy, and Subversion in Late Imperial Russia.* Cambridge, MA: Harvard University Press, 1998.
Robertson, Linda. *The Dream of Civilized Warfare: World War I Flying Aces and the American Imagination.* Minneapolis: University of Minnesota Press, 2003.
Rogger, Hans. *Russia in the Age of Modernization and Revolution, 1881–1917.* New York: Longman, 1983.
Rutherford, Ward. *The Tsar's War, 1914–1917: The Story of the Imperial Russian Army in the First World War.* 2nd ed. Cambridge: Sutton Publishing, 1992.
Ruzov, L. V. and IU. N. IAblochkin. *Gatchina. Istoricheskii ocherk.* Leningrad: Lenizdat, 1959.
Sanborn, Joshua. *Drafting the Russian Nation: Military Conscription, Total War, and Mass Politics, 1905–1925.* DeKalb, IL: Northern Illinois University Press, 2003.
Schimmelpenninck van der Oye, David and Bruce Menning, eds. *Reforming the Tsar's Army: Military Innovation in Imperial Russia from Peter the Great to the Revolution.* Washington, D.C.: Woodrow Wilson Center Press, 2004.
Shatsillo, K. F. *Rossiia pered pervoi mirovoi voinoi (vooruzhenye sily tsarizma v 1905–14 gg.).* Moscow: Nauka, 1974.
Showalter, Dennis E. "Army and Society in Imperial Germany: The Pains of Modernization." *Journal of Contemporary History* 18 (October 1983): 583–618.
Shumikhin, V. S. *Sovetskaia voennaia aviatsiia: 1917–1941.* Moscow: Nauka, 1986.
Shy, John. "The Cultural Approach to the History of War." *Journal of Military History* 57 (October 1993 supp.): 13–26, 59–62.
Sikorsky, I. *The Story of the Winged S.* New York: Dodd, Mead and Co., 1938.
Singer, Bayla. *Like Sex With Gods: An Unorthodox History of Flying.* College Station: Texas A&M University Press, 2003.
Snyder, Jack. *The Ideology of the Offensive. Military Decision Making and the Disasters of 1914.* Ithaca: Cornell University Press, 1984.
Spatarel', I. K. *Protiv chernogo barona.* Moscow: Voennoe Izdatel'stvo, 1967.
Spring, W. D. "Russia and the Franco-Russian Alliance, 1905–1914: Dependence or Interdependence?" *The Slavonic and East European Review* 66 (October 1988): 564–592.
Stariparloff, T. "The Russian Military Air Service up to the Revolution," *Air Power* 4 (December 1918): 324–337.
Steinberg, John. *All the Tsar's Men: Russia's General Staff and the Fate of the Empire, 1898–1914.* Baltimore: Johns Hopkins University Press, 2010.
Steinberg, John, et al., eds. *The Russo-Japanese War in Global Perspective: World War Zero.* Boston: Brill, 2005.
Sterret, James. *Soviet Air Force Theory, 1918–1945.* New York: Routledge, 2007.
Stokesbury, James L. *A Short History of Air Power.* New York: William Morrow and Co., 1986.
Stoll, Richard. "Steaming in the Dark? Rules, Rivals, and the British Navy, 1860–1913." *Journal of Conflict Resolution* 36 (June 1992): 263–283.
Stone, Norman. *The Eastern Front, 1914–1917.* New York: MacMillan, 1976.
Sukhomlinov, V. A. *Vospominaniia.* Berlin, 1924.
Swider, Raymond Jr. *Soviet Military Reform in the Twentieth Centuries: Three Case Studies.* Westport, CT: Greenwood Press, 1992.
Teitler, Gerhard. *The Genesis of the Professional Officers' Corps.* Beverly Hills: Sage, 1977.
Tkachev, V. *Russkii sokol.* Krasnodar, 1961.
Trunov, K. I. *Petr Nesterov.* Moscow: Sovetskaiia Rossiia, 1975.
Vagts, Alfred. *A History of Militarism: Romance and Realities of a Profession.* New York:

W.W. Norton, 1937.
Van Doorn, Jacques. *The Soldier and Social Change: Comparative Studies in the History and Sociology of the Military*. Beverly Hills: Sage, 1975.
Van Dyke, Carl. *Russian Imperial Military Doctrine and Education 1832–1914*. New York: Greenwood Press, 1990.
Veigelin, K. E. *Put' letchika Nesterova*. Moscow: Gos. Izdatel'stvo Oboronnoi Promyshl., 1939.
von Hagen, Mark. *Soldiers in the Proletarian Dictatorship: The Red Army and the Soviet Socialist State, 1917–1930*. Ithaca, NY: Cornell University Press, 1990.
von Laue, Theodore. *Sergei Witte and the Industrialization of Russia*. New York: Atheneum, 1973.
Walicki, Andrei. *A History of Russian Thought: From the Enlightenment to Marxism*. Stanford: Stanford University Press, 1982.
Weeks, Theodore. *Nation and State in late Imperial Russia: Nationalism and Russification on the Western Frontier, 1863–1914*. Dekalb, IL: Northern Illinois University Press, 1996.
Weiner, Martin. *English Culture and the Decline of the Industrial Spirit, 1850–1980*. New York: Cambridge University Press, 1981.
Whiting, Kenneth R. *Soviet Air Power*. Maxwell Air Force Base, Alabama, 1985.
Wildman, Alan K. *The End of the Russian Imperial Army: The Old Army and the Soldiers' Revolt (March–April 1917)*. Princeton: Princeton University Press, 1980.
Wilfong, Walter Thomas. "Rebuilding the Russian Army, 1905–1914. The Question of a Comprehensive Plan for National Defense." Ph.D. diss., Indiana University, 1977.
Wirtschafter, Elise K. *From Serf to Russian Soldier*. Princeton: Princeton University Press, 1990.
Wohl, Robert. *A Passion for Wings: Aviation and the Western Imagination, 1908–1918*. New Haven: Yale University Press, 1994.
———. *The Generation of 1914*. Cambridge, MA: Harvard University Press, 1979.
———. *The Spectacle of Flight: Aviation and the Western Imagination, 1920–1950*. New Haven. CT: Yale University Press, 2005.
Wohlforth, William C. "The Perception of Power: Russia in the pre-1914 Balance." *World Politics* 39 (April 1987): 353–381.
Wortman, Richard. *Scenarios of Power: Myth and Ceremony in Russian Monarchy from Peter the Great to the Abdication of Nicholas II*. abr. ed. Princeton: Princeton University Press, 2006.
Zaionchkovskii, A. M. *Podgotovka Rossii k imperialisticheskoi voine. Ocherki voennoi podgotovki i pervonachal'nykh planov. Po arkhivnym dokumentam*. Moscow: Gos. Voennoe Izdatel'stvo, 1926.
Zaionchkovskii, P. A. *The Russian Autocracy in Crisis, 1878–1882*. Gulf Breeze, FL: Academic International Press, 1979.
———. *Samoderzhavie i russkaia armiia na rubezhe XIX-XX stoletii, 1881–1903*. Moscow: Mysl', 1973.
Zalutskii, G. V. *Vydaiushchiesia russkie letchiki*. Moscow: Voennoe Izdatel'stvo, 1953.
Zhilin, A. P. "Bol'shaia programma po usileniiu russkoi armii," *Voenno-istoricheskii zhurnal* no. 7 (July 1974): 90–97.
Zhilin, P. A., ed. *Russkaia voennaia mysl' konets XIX- nachalo XX v*. Moscow: Nauka, 1982.
Zolotarov, V. A. "Russian Military Thought at the Turn of the Twentieth Century." *War and Society* 9:1 (May 1991): 51–62.

| Index

Studies in Modern European History

The monographs in this series focus upon aspects of the political, social, economic, cultural, and religious history of Europe from the Renaissance to the present. Emphasis is placed on the states of Western Europe, especially Great Britain, France, Italy, and Germany. While some of the volumes treat internal developments, others deal with movements such as liberalism, socialism, and industrialization, which transcend a particular country.

For further information about the series and submitting manuscripts, please contact:

Peter Lang Publishing, Inc.
Acquisitions Department
29 Broadway, 18th floor
New York, New York 10006

To order other books in this series, please contact our Customer Service Department:

(800) 770-LANG (within the U.S.)
(212) 647-7706 (outside the U.S.)
(212) 647-7707 FAX

or browse online by series at:

WWW.PETERLANG.COM

www.ingramcontent.com/pod-product-compliance
Ingram Content Group UK Ltd.
Pitfield, Milton Keynes, MK11 3LW, UK
UKHW021827100926
8556IPUK00003B/202

9 781433 114908